MODERN BIOLOGY

&

Visions of Humanity

© De Boeck & Larcier s.a., 2004
Éditions De Boeck Université
rue des Minimes 39, B-1000 Bruxelles
For the French edition and the co-ordination of the publication in all languages.
http://www.deboeck.com

© Multi-Science Publishing Co. Ltd., 5 Wates Way, Brentwood, CM15 9TB Essex, United Kingdom
For the English edition ISBN: 0 906522 30 7
http://www.multi-science.co.uk

The Italian version of the book is available with:
Centro Stampa d'Ateneo
http://www.uniroma1.it/centrostampa

Graphics and cover: www.cerise.be

Preface

The life sciences, which bring hope, and often solutions, in situations of pain and distress both in the developed world and in areas still struggling to achieve growth, are creating dramatic changes in the fields of knowledge and power. They offer humanity, that is to say doctors or engineers, processes that borrow from nature astonishing precision and… formidable efficiency. However, at the same time, the domestication of life's most intimate mechanisms has made them somewhat commonplace, putting our principles and our secular values to the test of performance and, let's not deny it, of our fantasies.

In this context, I attach great importance to public debates on the increasing impact of life sciences and biotechnological innovations. The event we are organising on 22 and 23 March 2004 in the city of Genoa, the cultural capital of Europe, will put high-calibre scientists in contact with personalities from the world of art and literature. What is, therefore, the place of life sciences in today's European cultures? How can we monitor scientific and technical progress and make sure that European citizens benefit from research developments? Good governance is impossible without responsible citizens. But is it possible to imagine citizenship without cultural awareness? It is inconceivable to discuss the repercussions of research without taking into account what science is able – or unable – to achieve. And can one imagine these discussions ignoring also the impact of such advances on culture, using the word here in its fullest sense? These crucial questions are the focus of our concerns and will be at the heart of debates in Genoa.

Philippe Busquin, European Research Commissioner

Contents

8

● ● ● The context: threat or opportunity?

We live at a time when the words science and biology trigger a strong social reaction combining fascination with fear – a time not unlike the Renaissance or Enlightenment. During those periods, geographical discoveries, new arts and sciences expanded the frontiers of knowledge. By contrast, our epoch brings a new understanding of the nature of life and of the cosmos through molecular biology and contemporary physics.

In their day, those earlier developments sent out unsettling messages about existing social orders and our perception of ourselves: Europe is not the centre of the world, the Earth is not the centre of the Universe, and there is no cosmic order that rules both the heavens and society. Darwin went on to show that we evolve by the same rules as the rest of the biological world. Now, DNA introduces another new perspective: all living beings, from bacteria to elephants, use basically the same genetic code. Our prized human genome is not very different from that of a small worm or a tiny weed. Bananas share 50% of their genes with us, and chimpanzees and bonobos more than 99%. Furthermore, 95% of biological diversity resides in the invisible microbial world.

Can we still cling to a traditional view of ourselves in light of these new data? Can contemporary religious and philosophical anthropologies still embrace a new vision of the living order, once genomes have been deciphered and the construction of life re-visited? The ultimate bastion of human uniqueness – the brain cortex – is the focus of much scientific inquiry, and recent discoveries question much of what we believe consciousness to be. Yet, at the same time, we have never had so much power to influence our future and make choices in our individual lives.

The classical Western questions: 'Where do we come from?' 'What is our destiny?' 'Why we are different from each other?' 'Are we free?', are now under the lens of scientific inquiry. How can freedom and individual choice still exist within the genetically determined framework apparently imposed on us? Is experimental biology turning into a full-scale ideology?

Besides challenging our view of ourselves, the discoveries of modern biology have a direct, often positive and sometimes spectacular impact on our existence. Many of them have improved our health, for example. We are witnessing the birth (not without a worldwide controversy) of a new agriculture based on the genetic design of novel properties in food staples. And new biodegradable materials are contributing to the evolution of a greener society.

At the same time, anti-globalisation activists argue that science and technology are as useful for political domination and supremacy as colonisation was in the past, while biotechnology seems to grant authorities alarming opportunities to invade individuals' private lives. But should we give up the obvious advances these new technologies offer, for fear of political and social abuses? This question is critical in a time when support for biological research depends so much on social acceptance and political viability. We have learned the hard way that no scientific endeavour can prosper in the short run when confronted by massive social rejection.

Today, modern biology is caught up in a three-sided argument between those who value the products and services that emerge from it, those who feel their beliefs challenged by its new vision of what humanity is, and those who question the whole approach of Western science from a political perspective. There are apparently few chances for open-minded discussion without pre-conceived conclusions. So the encounter on Modern Biology and Visions of Humanity in Genoa in March 2004 breaks new ground in setting up an informed and structured dialogue between scientists and researchers, and philosophers and politicians, artists, authors, educators and journalists.

●●● Genoa: the crossroads of European policy-making

In 1999, Philippe Busquin took over as the EU's European Research Commissioner. In April 2000 he set up the European Group on Life Sciences (EGLS), a body of 13 distinguished experts, to advise on the issues modern biology raises. Policy-makers have to confront a longstanding lack of mutual understanding between scientists and non-scientists, and to balance hard facts against public perceptions. Yet science advances within the context of human relations, not outside the social order, and widely diverging tendencies exist within the discipline itself. The European Commission conceived the pioneering Genoa encounter as a forum for influential figures from the arts and humanities to put forward contemporary views on phenomena such as the appearance of life and human nature to some of the leaders in modern biosciences[1].

The authors of the essays in this volume were invited to take part in the Modern Biology and Visions of Humanity event, and the order of the contributions follows the structure of the programme, reflecting four important aspects of the complex relationship between science and society: what are the benefits of science, how does it work, who does (or should) decide its uses, and how does science relate to the other main areas of human creativity? They are grouped under the headings:
- Life sciences and the belief in progress;
- The challenge and limitations of reductionism in the life sciences;
- Life sciences and democracy;
- Science and fiction: the cultural spin-offs from the life sciences.

1 The contribution by the organising committee for the Genova encounter – Victor de Lorenzo (Centro Nacional de Biotecnologia, Madrid), Halldor Stefansson (European Molecular Biology Laboratory, Heidelberg) and Etienne Magnien (European Commission) – is gratefully acknowledged.

●●● An open debate: towards a second Enlightenment?

Science and Society meetings are not unusual. Yet recent developments such as the ability to produce DNA sequence in mass (with the final outcome of the human genome), genetic screening procedures, the rise of bio-informatics and other revolutionary developments in the omics sciences (genomics, proteomics, metabolomics etc) set new precedents that impact on our lives more than ever. This demands a fresh discussion, with thinkers and individuals from all sides taking part. What is at stake is a new view of ourselves that will either add momentum or divert the development of life sciences research in Europe. Without the full support of its citizens, the dream of converting our continent into a consensual knowledge-based society and economy by the end of this decade will simply not become reality. Throughout their history, the European people have produced artists and writers who captured the spirit of the times and the winds of change much earlier than mainstream society. Giotto broke away from the static, naive depictions of reality which dominated the Middle Ages. Zola adopted narrative techniques taken from experimental sciences and his stories heralded the convulsions of the turn of the century. Mid- and late-19th century Impressionism was a formidable vehicle for exposing the hidden energy behind the material world, which contemporary physics research was revealing. Cinema was created in a moment of general optimism about the value of science and progress. Artists and writers, as well as scientists and thinkers, thus have an essential role in the debate on the rise of biology as the paradigm of knowledge of our time. Perhaps this unique combination of science, humanistic thought, political action and art can bring further enlightenment if this learning process remains inclusive of all citizens.

Life Sciences and the Belief in Progress

Ever since the Enlightenment, Western science has tended to take a positive, optimistic approach to the development of human society.

Above all, modern science has taken it for granted that the knowledge and understanding derived from science and technology will benefit humanity, and that they must be pursued at all costs, as key to improving the human condition. As industrial societies around the world have become more secular, this belief has fuelled the expansion of modern civilisation and spread patterns of development across societies and cultures.

Since the dawn of Western philosophy, people have also questioned whether logic and rational thought is different – or superior – to emotions. Neurobiologists and cognitive scientists have recently challenged this conventionally dualistic approach, while anthropologists have demonstrated how our own judgements, tastes and values tend to appear to us perfectly rational, yet those of people outside our cultures seem deplorably irrational.

Thus, alongside the largely dominant, positive view of Western science, other opinions of social history, ranging from the moderately sceptical to the downright pessimistic, now compete for public attention and popular support. Most recently, sceptics have raised questions not only about the use, but also about the risk of abuse of the know-how and technologies growing out of the life sciences. How can we assess past, present, and future benefits from the extraordinary discoveries and innovations that have been made in recent times within the life sciences? What is the relationship between social change and social progress? Who benefits from progress in the life sciences? Can such progress be harnessed to help diminish the appalling inequalities that exist around the world, or are the life sciences too much a part of the system that brought these inequalities about and continues to maintain them?

SCIENCE AND PROGRESS

Axel Kahn

Professor Axel Kahn (MD, PhD) was the European Group on Life Sciences' president from 2000, when the Group started, until the end of 2001. He works at Institut Cochin (FR) where he is head of the genetic and molecular physiology and pathology research unit. His research focuses on genetics, gene therapy, cancer, development and physiology for which he has received several honorary distinctions. He has written four books on modern biology, genes, and ethics, as well as some 400 papers in international journals. For nine years, he presided over the Biomolecular Engineering Committee of the French Ministry of Agriculture. He is currently a member of the French National Consultative Ethics Committee.

In the 17th century, Europe invented what we might call the Western model of development and society, based on knowledge and Progress. It is this model that today constitutes the standard for our globalised world. The principles of this model can be summarised in three quotations. "Knowledge is power", Englishman Francis Bacon wrote at the start of the 17th century. "Man's role is to make himself the master and owner of nature", Frenchman René Descartes added a few years later. Or, in the words of French philosopher and mathematician Blaise Pascal: "All mankind for so many centuries until now can be viewed as a single ever-living and continuously learning man." From this we can conclude that the power conferred by infinitely cumulative knowledge leads to a constantly growing technical power. It is this power that forms the basis of the economic expansion of the nations that have signed up to this model.

The Enlightenment philosophers would add, in the 18th century, that these mechanisms guarantee human happiness.

● ● ● Louise Michel and the Kiev Couple

Revolutionary Louise Michel's statements to the court, which condemned her to be deported to New Caledonia for her active part in the Paris Commune, make moving reading. This primary school teacher had devoted her entire life to educating underprivileged children in the poor districts of Paris, before joining the combat on the Parisian barricades. Louise Michel challenged her judges with her hope of a world where people, having accessed knowledge, will break their chains and pave the way for a new world of solidarity. This conviction has been shared by republicans and militants in the social struggle who defined themselves as "men of progress". Indeed, the term "progressist" itself was to become synonymous with partisan among these combats – a member of the political left. It is this ideal which was originally symbolised by the Kiev couple depicted in this monumental sculpture in the style of Socialist Realism. I remember my encounter with these young stoneworkers, the man with his energetic, determined face, the woman haughty and beautiful, her eyes set towards a radiant future. These were the dying days of the USSR, Gorbatchev was still in power, execrated by all his entire fellow-citizens. In those days, the shops and people's plates were nearly empty, and I ate little more than bacon and onions for an entire week. And I asked myself what this couple really saw, from their promontory overlooking the Dnieper, what had become of Louise Michel's dream. On the one hand, Progress has achieved certain of its objectives, and education has undoubtedly advanced throughout the world. Very few children die in infancy – a very real fate a hundred years ago – almost no women die in childbirth, and they have also acquired control of their fertility. In developed countries, life expectancy at birth has practically doubled in 100 years, even if it has scarcely advanced in other disadvantaged regions such as black Africa. The physically exhausting aspect of work has diminished, too often to be replaced by a sometimes intolerable level of psychological pressure. Citizens can engage in dialogue 20,000 kilometres apart, opening the way to globally networked communities. Despite this, right across the world, the same channels of our communication and information society are constantly carrying calls for exclusion, hate, murder … and it is far from certain that understanding between Neuilly and Saint-Denis, or between Long Island and the Bronx, is any better than it used to be. In the not too distant future, we shall probably see man step on to new planets, communicate by text, sound and image worldwide, and beyond. A combination of informatics, microelectronics, robotics and micro-mechanics will permit us to increase, in proportions unimaginable until only recently, the conjoined powers of the human hand and mind, and to repair ever more effectively bodies damaged by accident and disease. Our various maladies will be increasingly well understood, predicted and mastered thanks to the expected progress in the exploitation of data from genome, and in particular human genome, research. Growing mastery of cell and tissue regeneration will enable us, to a certain extent, to repair the irreparable ravages of the years, and many children born today will live to see the 22nd century. On the other hand, what our young Soviets contemplate, in the foreground, is a dried up Lake Aral and the ecological disaster of an ideologised industrialisation, the dramatic increase in the alcoholism that Communism never managed to stem, mafias, violence, inequality, falling living standards, etc. Thousands of miles away, in the same direction, lies Manilla, a major, modern, westernised city, where hundreds of thousands of people scrounge a living from gigantic open-air refuse tips piled high with the waste of a well-to-do population. During particularly heavy rains, the contents of these waste mountains slip down, burying and killing hundreds of people living cheek by jowl with them.

This episode illustrates the staggering growth of inequalities which make the misery even more unbearable. The tensions and conflicts arising from the impudence of the daily flaunting of a debauchery of wealth and well-being in the face of impoverishment, in our globally communicating world, constitute a threat to everyone, including the well-off. One day or another, the unacceptable will no longer be accepted. How, for example, do we qualify the incredible growth, after a century of scientific and technical progress, of the worst of inequalities, the fact of being unequal in the face of sickness and death? Life expectancy in many African countries today is 30 years lower than in wealthy countries, whereas there was little difference in the 19th century. Thirty-six million of the world's 40 million people AIDS sufferers in 2003 live in poor countries, 30 million of them in Africa. The fact is that just 5% of the funds allotted for combating this malady are devoted to them, with the remaining 95% going to the small group of sick people in the wealthy countries. Here, too, research has made enormous strides, as several HIV-positive people live despite the odds, in the USA and Europe. However, only minimal efforts are being made to prevent other even more fearful scourges, such as most of the parasite-borne diseases. Indeed, for those affecting only Southern Hemisphere countries, these efforts are in decline. Our overall assessment of the 20th century is therefore a mixed one. On the one hand, there has been a gigantic growth in knowledge, technology, and wealth production; on the other, two world wars fought with the latest technology, genocides, the Hiroshima and Nagasaki bombs, the Tchernobyl and Bhopal accidents, pollution, asbestos, contaminated blood and mad cow scandals and, as we have just seen, the explosion of inequality. Nor have the first years of the 21st century done anything to reassure us. Everything that has been the cause of concern and revolt during the previous century continues. Huge amounts of scientific and technological energy continue to be poured into developing ever-more effective armaments. That said, it was with rudimentary fake arms that on September 11, 2001, a handful of young men highjacked the aircraft which were to destroy the two proud South Manhattan towers. These men came from the wealthiest and best educated classes in their respective countries. They had studied at the world's best universities. Even so, despite educational, scientific and technological process, it is in the name of the most disreputable of passions that they killed men and women from the world over who were at work in the World Trade Centre. Visibly, access to knowledge was not enough to nurture in them a love of the Other and of freedom. Decidedly, there is something wrong in our optimistic belief in Progress, which explains why hardly anyone shares it today. Looked at objectively, the threats confronting the world today are no more fearful than at the start of previous centuries. In 1700, Europe was suffering the effects of the wars of French King Louis XIV, the War of the Spanish Succession, the terrible end of his reign. In 1800, fresh from the convulsions of the French Revolution, Europe was about to tear itself apart in the Napoleonic wars, followed in the second half of the century by the Franco-Prussian war, and by the terrible War of Secession on the other side of the Atlantic. In 1900, the stage was being set for the massacres of the World War, the Armenian genocide and the Russian Revolution, preluding the cataclysm of fascism and the paroxysm of the 1939-45 World War. Yes, the horizon at the start of our new century is dark, but no more so than in the past. What is new is the general decline in the hope that Progress, ultimately, inevitably, will grant mankind mastery of a destiny finally freed from excessive violence and able to promote the happiness and fulfilment of everyone. In fact, the dominant image is of a future that is at best uncertain, and at worst severely compromised. This feeling is fuelled, of course, by the social difficulties which do not spare rich countries, the loss of the advantages gained in hard struggle, inequality, terrorism and the chequered success story of technology in terms of its effects on the conditions for human happiness and fulfilment. Not only this – we are now witnessing a theoreticisation of the unavoidable failure of Progress to civilise and pacify the world. An absolute contradiction exists between the American-European model of development that is being imposed on the entire world, both driving force and consequence of the cultural and economic globalisation marking the turn of the century, and the obvious impossibility of achieving the objectives inherent in this model. Every people in the world, both those hostile to the United States and their strongest allies, wants to achieve American standards of living. Modern means of communication are being used, for obvious reasons of economic competitiveness, to promote this model. Of course, as we know, it is simply impossible to generalise the "American way of life". To do so would be to sound the death knell of our planet. For example, every US citizen consumes an average 5 000m^3 of renewable water annually. Were the world's reserves of this precious liquid to be equally distributed, each human being on earth would receive at most 2 000m^3 each. In addition, the reserves in most Southern countries are less than 500, and in certain cases 200m^3 per inhabitant. America's 300 million or so inhabitants, one 20th of the world's population, consume a quarter of the world's energy. In other words, so-called "globalisation" is ending up creating desires that are both strong but impossible to meet. The consequences of such a contradiction are well known: frustration, revolt, violence…. Faced with this situation, two types of discourse are appearing, one vaunting the merits of a "flight into the future", the other calling for a paradigm change. For the former, technology will always find solutions to the problems it causes. For the latter, we need to ask ourselves urgently what really are the most favourable conditions for general personal fulfilment, today and tomorrow, and to redirect our use of human knowledge and activities accordingly.

● ● ● Knowledge, Progress and Freedom

Science is the enterprise of the mind, aimed at acceding to knowledge and, in particular, that of the laws of nature. Its tools are reason, logic, observation and, in many cases, measuring. The ability to gather knowledge is a universal one. In fact, one of the anthropological features of man, who could be defined as a curious primate, is possessing the means to understand and to pass on to his successors what he has learned. The rational method of accessing knowledge, based on observation – interpretation – deduction – induction – confirmation, is historically more recent, deriving from Babylonian mathematics and the Greek logos. In Jules Michelet's words: "Everything stems from the invention of reason in Greece. The river started flowing, never to stop again, from Solon[1] to Papinius[2], from Socrates to Descartes and from Archimedes to Newton." Greece did indeed invent science as we practise it today, but not, of course, knowledge. The Egyptians and the Mesopotamian civilisations already possessed a considerable body of knowledge, in astrology and at times also in physiology and medicine. Trepanning was practised in ancient Egypt. What was lacking, however, was any systematised method for achieving a generalised understanding of the laws of knowledge. Logic had no part to play. Knowledge was something given, received, handed on. The first proof of a modern scientific approach comes from Babylonia, 40 centuries ago, in the form of clay tablets bearing algebraic equations, perhaps the exercises of Mesopotamian schoolchildren. However, discourse on the rational path of scientific thinking derives effectively from the handful of Greek communities which, from Thales to Zeno of Elea, Leucippus and Democritus, were to develop the principles of logic (invention of

---

1 The lawmaker of 6th century BC Athens (note by the author).
2 The 2nd century AD Roman poet (note by the author).

the syllogism), leading to a method of access to knowledge and development of the sciences. This having been said, Progress is not a Greek invention. Its civilisation lacks any concept of the possible improvement of society by means of knowledge: the Greeks had no intention at all of improving society by means of scientific progress, for several reasons. First of all, certain Greek authors, such as Plato, believed knowledge to be something pre-existent which simply had to be rediscovered. And whilst, for the great authors of Antiquity, Greek civilisation is certainly an improvement on the barbarian world preceding it, it also represents an unsurpassable summit, and there is no question of dreaming of its progress.

However, whilst the Greece that invented science did not think in terms of Progress, it is to it that we owe one of the roots of the 18th Progressism, in the form of a question that the Greeks had posed since pre-Socratic times, but in particular with Socrates, Plato and Aristotle: what is the relationship between the various categories enabling us to attribute values to things and to people: True and False, Good and Evil, Beautiful and Ugly? More particularly, what is the relationship between the True and the Good? Two theses clash here. The first is that of Socrates, expressed by his pupil Plato, for whom a person with knowledge is unable to behave badly, such behaviour being explicable solely by a lack of knowledge. The second is that of the Sophists, with *Protagoras* or *Georgias*, as set out superbly in the philosophical dialogues that Plato puts into the mouth of his master Socrates. In order to clearly denote the difference between technical knowledge, resulting from research for the True, and the quest for Good, understood in the sense of public virtues, Protagoras has Socrates comment that[3]: "The Athenians, and mankind in general, allow but a few people to give advice when it comes to the art of building or any

---

3 PLATO, *Protagoras*, pp. 322-323.

other mechanical art. If anyone else from outside this little group intervenes to express an opinion, they do not tolerate it, and they are right. But when one is deliberating on politics, and everything is based on justice and temperance, they are right to admit everyone to the discussion, because it is important that everyone be involved in civic virtue, otherwise there is no city. Here, Socrates, lies reason of the difference between science and virtue." In *Protagoras'* epilogue to his account of the myth of Prometheus, we see human beings endowed with the divine fire, arts and technologies, but under threat of extinction because of their inability to agree among themselves and to unite against the wild beasts, instead of quarrelling and killing each other all the time. To save them, something that science alone cannot achieve, Zeus has to task his messenger Hermes with endowing them all with justice, prudence, and a sense of solidarity. What the sophist Protagoras is saying here is that whilst science and technology are certainly essential for human existence, they are not enough. Also needed are *dike* and *aidas*, justice and solidarity, that is, the sense of the other, and a consciousness of the essential value of the Other. In this conflict, Plato and Socrates are visibly not the ancestors of Progressism, since the concept of the improvement of society by progress was unthinkable then. On the other hand, they are the precursors of an idea of a determinism of science and of knowledge, the notion according to which man is determined by essence to make good use of his knowledge: a man possessing knowledge cannot but chose good, as this is what he is made for. Protagoras' thinking, on the other hand, was to be picked up several centuries later by Rabelais, in his too often misquoted maxim: *"Knowledge without conscience is the ruin of the soul"*, stating hereby that such ruin is possible since having knowledge does not necessarily imply having a conscience. The other root of the idea of Progress and Progressism is the recognition, by Bacon and then Descartes, that knowledge represents the power to become both master and owner of nature. Condorcet and the other

23

eulogists of progress of the succeeding centuries, would take over both this concept of the power conferred by science and knowledge, and Plato's optimism as to the moral nature of knowledge. Science sets itself the objective of providing a plausible and rational explanation of the world, and in so doing aspires to truth. For them, truth is unquestionably a positive value, its contrary being error or untruth, to which it is not possible to assign any virtues. The result is a sort of sophism which has exacted a terrible toll: "Truth is preferable to lies, light to dark." Now the pursuit of truth and light is the purpose of science. Science is therefore inherently good. Scientific knowledge gives access to power which, in turn, can only benefit from the positive moral connotation accorded to truth and scientific knowledge. The fundamental error of such thinking is easy to demonstrate, as well as being amply confirmed by experience. Nobody doubts that knowledge gives power. The 20th century testifies to the unquestioned and fabulous progress brought about by the growth of knowledge in biology, medicine, technology and many other fields. On the other hand, what is there to say that such knowledge will necessarily be used for the benefit of mankind? Indeed, such an affirmation sounds wrong, because if the only way that the power deriving from knowledge could be used were for the benefit of humanity, this would signify that man is determined by nature to make only one use of it. In this case, we would cease to be responsible for it, as we are responsible only for what we are free to do or not do. The concept of a good deed has meaning only when faced with man's singular capacity to do evil. Good and Evil are both consubstantial with the idea of freedom, whatever the uncertainties attached to this word, and this freedom is the precondition for responsibility. It is because knowledge – and the power derived from it – in no way prescribes us to use it to either the benefit or detriment of our fellow humans, for life or for death, for freedom or constraint, that we are individually and collectively responsible.

What are the preconditions for access to responsibility? Clearly, knowledge is one of these: we are not responsible for what we do not know, and we can be responsible only if we know what is at stake. In this sense, science is clearly an essential factor in any ability to exercise freedom: by enlightening us on the nature and consequences of the alternatives of a choice, where these involve scientific and technological questions, it enables us to assess the issues at stake. In this way, science becomes one of the determinants of choice, leaving to personal feeling – or what we sense to be this – the possibility of manifesting itself with full knowledge of the facts.

A second precondition is man's aptitude for moral judgement, recognition of the value of the Other, and the ability to assess the consequences for him of the choices he makes and the deeds he undertakes.

A third component of responsibility, and probably the hardest to attain, is learning the desire for freedom. I use here the term "desire for freedom" to clearly mark my genuine uncertainty as to what this really is. There is nothing obvious here, as man is not genetically predetermined to desire freedom. In fact, the reason why *Homo sapiens* is so easy to educate lies in his astonishing capacity, on the biological level, for "domestication". Education is based, at the outset, on the sense of imitation. Man is probably the most gifted living being of all when it comes to imitating others. This also makes him the most susceptible to epidemic infection by one of the scourges of humanity, the virus of ideology, and of the reproduction of archetypes. Knowledge alone is not enough to combat this fanaticism. Indeed, knowledge can serve to fuel fanaticism. This obvious fact confers on every teacher the duty of doing more than just passing on knowledge. Of course, learning about things and being remains essential, but it is also necessary to learn to learn, and to learn to doubt. Familiarising young people with the legitimacy of doubt, making them aware of the need for scepticism and dissent is another mission of the educational system. But even this is not enough. Families, schools and universities would all fall short of their task is they did not make every effort to inculcate into tomorrow's citizens a sense of their responsibilities in using the power derived from acquired knowledge, as a function of the values that they adhere to, certain of which have a claim on universality since they are based on respect for the other. We have still a way to go to realise such an ideal. Much-repeated affirmations, such as *"You cannot stop Progress"*, are eloquent testimony of the permanence of a deterministic vision of Progress and its injunctions, from which there is no escape. The progress of scientific and technological education does indeed offer a way for man to exercise his freedom. But there is nothing to say that he will use it to the benefit of humanity. Examples of this abound, but I would like to illustrate it with the emblematic figure of Fritz Haber. This great German chemist was a benefactor of humanity, a Nobel prizewinner in chemistry, in 1918, just after the Great War. It is he who invented the industrial synthesis of ammoniac. The resulting nitrogenised fertilisers permitted a considerable increase in farm yields at the time when traditional sources of organic fertilisers were running out. Tens of millions of people probably escaped famine thanks to him. But Haber was also a German patriot entirely dedicated to the success of the Reich's armies. Seconded to the War Ministry from the start of hostilities, in 1916 he became director of the army's chemicals services, responsible for warfare gases. It is he who, in 1915, supervised the first chlorine attacks, close to the Belgian town of Ypres, of which the Franco-Senegalese troops were the victims. From 1916 onwards, Haber shared full responsibility for the development of even more fearsome gases, phosgene, then yperite. Right at the end of the war, Haber and his team developed Zyklon B, first manufactured as an insecticide to meet the conditions imposed by the armistice, and subsequently used in the Nazi gas chambers. Fritz Haber was a Jew and resigned his university post in 1933, following the promulgation of Chancellor Hitler's first racial laws. Dying shortly after, he did not witness the use of his latest invention to eliminate his fellow Jews. In 1915, after the German troops had used chlorine, on Fritz Haber's advice, his wife Clara, herself a talented chemist, reprimanded him severely. For her, a science that sets out to kill has deviated from its purpose. Fritz Haber turned a deaf ear. That evening, Clara seized his service pistol and shot herself dead. Fritz and Clara – two faces of science, two expressions of a scientist's responsibility. Fritz Haber's drama and the ambiguity of his personality provide a striking example of the total dissociation of the scientific value of a discovery and its social and moral consequences.

The scientific pseudo-theories forming the basis for the Reich's policy of racial hygiene were more or less approved, or more or less tolerated, by a large portion of the German biologist community, including some of the most outstanding names. In almost every country, genetic companies were then called eugenics companies, by referring to the eugenicist theory which concluded that it is necessary to improve human lineages. The implementation of eugenicist policies was to lead to the forced sterilisation of hundreds of thousands of people throughout the world. This ideological effervescence is one of the consequences of the progress of biology in the 19th and 20th centuries, of Jean-Baptiste Lamarck and Charles Darwin's theories of evolution, and of Gregor Mendel's discovery of the laws of genetics in 1865, more particularly their rediscovery from 1900 onwards. This scientific progress was considerable, constituting the pillars of modern biology. At the same time, however, it played a significant role in the social and political events of one of the darkest periods in human history, providing a striking demonstration of the fact that, as Protagoras stated, and contrary to the Socrato-Platonic position, "Good cannot be reduced to Truth".

All in all, ownership of knowledge and the mastery of technology, and the power they confer, always end up forcing mankind to make choices, the terms of which cannot be reduced to their scientific dimension. Such an observation in no ways detracts from the legitimacy of science or the value of truth, but rather states their purpose with greater precision.

BEYOND ENLIGHTENMENT:
THE RISE OF A CULTURE OF LIFE

Karin Knorr Cetina

Karin Knorr has been a sociology professor at Bielefeld University (DE) for the past 20 years, except for brief stints as a visiting professor at Princeton University (US). In addition to her three degrees, Professor Knorr has received several honours, including Vienna University's Fellowship of the Gifted, and she was a Ford Foundation Post-Doctoral Fellow. She is on the editorial board of several publications, including *Sociological Forum* and *Zeitschrift für Wissenschaftsforschung*. She has also published numerous papers and books, including *Epistemic cultures. The cultures of knowledge societies* (1999, Harvard).

Enlightenment ideals celebrated the human – human rationality, human potential, human freedom, equality and justice. These ideas included the belief in science but as a means through which improvements in human existence were to be achieved. Enlightenment ideals[1] lie at the roots of the belief in the perfectibility of and salvation by society that provide the moral underpinning for the human and social sciences. What the biological sciences promise, against these scenarios, is the *perfectibility of life*. The human, humanity and humanism have been integrative concepts inspiring intellectual elites and whole disciplines for centuries. Today, the notion of life replaces the notion of the human as a concept that bridges developments in several sciences and in practical discourses. "Life" stands for an open-ended series of biological, psychological,

---

1 European enlightenment dates back to the Renaissance and Reformation, fought the dogmatism of the church, and replaced it with the idea that humanness entails the capacity of reason.

economic, even phenomenological significations and processes. What it does not stand for is the further expansion of enlightenment ideals of human reason and social salvation. My argument is that we experience a turn to a "culture of life" in a broad and encompassing sense comparable to the broad sense in which human-centred ideas have dominated our thinking in the past. This development coincides with historical changes through which the culture of the human and the social that was based on enlightenment ideals empties out into a postsocial era. Life-centred ideas arise at a historical moment when the collective imagination seeks new stimulation. I also claim that these ideas are on one level deeply rooted in the biological sciences from which they draw motivation. But they are also in a more general sense embedded in developments that occur in other fields and practical areas. Thus a culture of life does not turn only on biology but is in fact nourished and sustained by a large number of processes and transitions. I will also briefly argue, based on speech act theory, that the concept of a promise that underlies a culture of life entails shifts in responsibility and temporal orientation that need to be brought out. Some of these shifts are apparent in political and other debates that I will mention at the end.

⚬ ● ● From a social Time to a postsocial Era

Sociality is very likely a permanent feature of human life. But the forms of sociality are nonetheless changing, and the regions of social structuring may expand or contract in conjunction with concrete historical developments. Modernity has often been associated with the collapse of community and tradition and the onset of individualisation. Central to our experience today are similar retractions of social principles in different regions of social life. These are not usually discussed together, and they do not have the same roots. But they work together in emptying out previous categories of social imagination, and in creating the space in which a postsocial imagination can take hold[2].

The first region of expansion of social principles during the course of the 19th century and throughout the early decades of the 20th was that of social policies, and this was linked to the rise of the nation state. Social policies as we know them today derive from what Wittrock and ÓWagner call the "nationalisation of social responsibility" – the formulation of social rights alongside individual rights and the positing of the state as the "natural container" and provider of labour regulations, pension and welfare provisions, unemployment insurance, and public education. A second region of expansion, connected to the first, was that of social thinking and social imagination. A corollary of the institutionalisation of social policies was new concepts of the forces that determine human destiny: they were now more likely to be thought of as impersonal, social forces. Rather than assuming the automatic adaptation of individuals to changing environmental conditions, Rabinbach argues, these ideas focused on the prevailing imbalances and their social causes: the social causes of occupational accidents would be an example. Sociology played an important role in bringing about the shift in mentality through which individuals came to be seen as the bearers of the individual costs of collective structures. A third area of expansion of social principles and structures was that of social organisation. The rise of the nation state implied the rise of bureaucratic institutions. The growth of industrial production brought with it the emergence of the factory and the modern corporation. Industrial, nation-state societies are unthinkable without complex modern organisations. Complex organisations are localised social arrangements serving to manage work and services in collective frameworks by social structural means.

---

2 For a fuller discussion, see KNORR CETINA, K., « Sociality with Objects. Social Relations in Postsocial Knowledge Societies », *Theory, Culture & Society*, 14(4), 1997, pp. 1-30.

A fourth area of expansion was that of social structure. The class differentiation of modern society is itself an outgrowth of the industrial revolution and its political consequences as well as of processes of social and political measurement and categorisation.

It is central to our experience today however that these expansions of social principles and of socially constituted environments have come to a halt. In many European countries and in the United States the welfare state, with all its manifestations of social policy and collective insurance against individual disaster, is in the process of being "overhauled"; some would say "dismantled". Thatcherism in Britain and "neo-liberalism" in general could be viewed, according to Urry, as a partially successful attempt to contest some of the social rights acquired in the previous half century. Social explanations and social thinking run up against, among other things, biological and economic accounts of human behaviour against which they have to prove their worth. The mobilisation of a social imagination was an attempt to identify the collective basis for individuals' predicaments and dispositions to react. This collective basis is now more likely to be found in the similarity of the genetic make-up of socially unrelated members of the population. Social structures also seem to be losing some of their hold. When complex organisations are dissolved into networks of smaller independent profit centres, some of the layered structural depth of the hierarchically organised social systems that organisations used to represent gets lost on the way. When person-provided services are replaced by automated electronic services, no social structures at all need to be in place – only electronic information structures. The expansion of societies to global societies does not imply, it appears, further expansions of social complexity. The installation of a "world-society" would seem to be feasible with the help of individuals and social microstructures, and perhaps becomes plausible only in relation to such structures.

●●● The Rise of a subject-centred Imagination

One of the most important elements in the development described so far may well be the loss of a social imagination, the slow erosion of the belief in salvation by society. The expansion of a social imagination involved, from the beginning, not only the idea of impersonal social forces affecting the individual but also the notion of universal human perfection through society. This idea was put forward by Rousseau and enlightenment thinkers such as Condorcet, who announced the possibility of an ever more rapid progress towards a perfect form of human society marked by "the abolition of inequality between nations, the progress of equality within each nation, and the true perfection of mankind". The notion is also epitomised by Marx´ vision of a socialist age, which he thought would begin once capitalism reached its peak and collapsed under its own self-created contradictions. The collapse of Marxism as a creed signifies the end of the belief in salvation by society; the end of a social imagination that transposed itself into what P. Drucker called a "secular religion".

Marxism also failed in practice, but its failure as a creed that supports the belief in society may be vastly more consequential. The question it raises is what replaces the social imagination that it sustained. In the sciences focused on the human world, the answer is plain: the fantasised unit is now the individual rather than society. This focus on the individual prepares the ground, I maintain, for the emergence of a notion of life that can be seen as a link between the human and the natural sciences.

The claim that the social imagination of the past is being superseded by individualising views can be linked to several developments. First, even from within the state itself, voices and slogans have emerged which advocate individual self-reliance in regard to personal welfare and non-governmental avenues to the achievement of collective goals. One illustration of this is the model of a de-institutionalised welfare state and a socialism that reinstitutes individual responsibility while curtailing the possibilities for benefit-seeking and social rights. Second, just as a social mentality was elaborated and extended by social science, so individualising ideas are unfolded by particular disciplinary traditions. One example is the rise of rational choice theories, which draw on concepts of utility maximisation in economics that have been imported into other sciences. These theories empower the individual as the unit that seeks information, calculates behavioural outcomes, and through these mechanisms, engineers his or her fate. The exaggerated emphasis in these models on instrumental reason and complete information, and the attempt to translate collective and cooperative choices into individual utilities may be "phantasmatic" (not warranted by data or plausible argument) from an empirical perspective, but these phantasms are also the ones that empower subjectivity thinking and cast doubt on social thinking. Theories of identity and identity politics, of the self and subjectivity provide other examples of such trends. Third, subjectivity thinking and subjectivity imagination is manifest in the vast numbers of self-help books and manuals that counsel individuals on self-improvement and engage them in the discovery of their own selves.

We can interpret this literature to suggest that more is involved in subjectivity thinking than the onslaught of a new wave of individualisation. The self-help literature consistently asserts an individual's right and obligation to make a strong commitment to him/herself. But it also affirms that it is a persons "life" that should be improved; the goal of the benefits that are to flow from the commitment to oneself (and from other strategies) is individual life-enhancement. Subjectivity thinking in the social and psychological sciences and in practical reasoning includes the notion of individual life as relevant to self-oriented behaviour. Second, the notion of the individual subject has itself become redefined in current thinking, and is now less divided from nature or non-human objects than it was before. Enlightenment thinkers drew the "circle of humanity" tightly, as Seidler argued, defining the subject in terms of reason that underpinned the subject's capacity to exercise agency. Sciences today (psychoanalysis, cognitive- and evolutionary psychology, the psychology of emotions, behavioural economics and biology are examples) draw the circle much more widely as they make claims about the unconscious and emotional sides of the individual, about human decision biases, and about human intelligence. As behavioural economists such as Richard Thaler have suggested, as assumptions about rationality give way to research into human cognition, *homo sapiens* loses IQ and gains visceral definition. The remarkable rise of subjectivity thinking thus harbours tendencies that bridge the gap between the human and the natural sciences. One tendency is the orientation to concepts such as the notion of life that serve as anchoring concepts in both areas of investigation. The second tendency is the orientation to research and explanations that assimilate humans with other forms of life. Both tendencies play into a more general culture of life.

Culture of Life

I now want to address life-centred ideas more directly, arguing that they span several disciplines and practical areas. The notion of life serves as a metaphor and anchoring concept that illustrates a cultural turn to nature and how it replaces the culture of the human and the social. What has become thinkable today in a break with enlightenment ideals is not the perfectibility of human society by societal means or the cognitive and ethical perfectibility of the human but the perfectibility of life – through life enhancement on the individual level, but also through the biopolitics of populations, through the protection and reflexive manipulation of nature, through the idea of intergenerational (rather than distributional) justice.

One massive source of fantasies that fuel a culture of life and challenge traditional humanism is the biological sciences themselves, which warrant special attention. The biological sciences produce a stream of research that inspires imaginative elaborations of the human individual as enriched by genetic, biological, and bio-technological supplements and upgrades. These ideas relate to the enhancement of life through pre-implantation genetic diagnosis and screening, germ-line engineering (genetic changes that can be passed down to an individual's offspring), psychotropic drugs that improve emotions and self-esteem, biotechnological means of enhancing the life span, and human cloning. Serious research money goes into the halting and reversing of ageing processes that give us the option of extending our physical and intellectual capacities beyond present levels. Much fantasising and some research money goes into the improvement of cloning techniques, despite its condemnation by many segments of the population and some governments, and despite concerns about a "dehumanised" future and the rights many think we have to a unique or unknown genome. The very debates surrounding the biological sciences bring into focus the perfectibility of individual life that the biological sciences promise.

The biological sciences are central to life-centred thinking, but the respective ideas also pervade other fields. The extent to which this happens is illustrated by those areas that see human beings as on the verge of being transformed into *cyborgs*[3], *posthumans*[4] or *trans-humans*[5]. These creatures are human descendants that have been improved upon by bioengineering in combination with nanotechnology, the information sciences and cognitive research (the "NBIC" group of

---

3 HARAWAY, D., *Simians, Cyborgs, and Women*, New York, Routledge, 1991.
4 FUKUYAMA, F., *Our Posthuman Future: Consequences of the Biotechnology Revolution*, New York, Farrar, Straus and Giroux, 2002.
5 BAARD, E., « Cyborg Liberation Front. Inside the Movement for Posthuman Rights », *The Village Voice*, 30 juillet 2003.

sciences)[6]. The NBIC sciences converge in investigating and developing devices that enhance, or "augment" biological human nature. The challenge they pose is that they put into question sharp distinctions between humans and machines, and that they appeal to the plasticity of the organic and the technical. NBIC research creates interfaces between these categories that blur the human-non-human distinction. While most technologies currently implanted in humans (for example neural implants for Parkinson's disease) or used as replacement parts for malfunctioning organs and the like have medical functions, it appears likely and is expected by some authorities (e.g. the United States National Science Foundation and Department of Commerce) that the convergence of the NBIC sciences will create opportunities to enhance normal human performance and will be used for these purposes. Performance enhancements might comprise an expanded memory capacity, implanted links for direct access to telecommunication networks, much faster thinking-speed or the capacity to see in the infrared and ultraviolet wavelength. Such "improvements" are not within immediate reach, but they may well be realised in some way in the not too distant future. Overcoming genetic diseases (such as asthma, cystic fibrosis, epilepsy, muscular dystrophy, Parkinson's disease, sickle cell anaemia), congenital birth defects (resulting in Down syndrome, spina bifida, and cerebral palsy) and many other causes of human suffering, and overcoming the normal predicaments of old age and perhaps death, also provide motivation for the creation and testing of relevant devices.

Developments of this sort raise many questions about the rights extended to trans-humans and the ethics expected from them. Will those with a higher percentage of prosthetic parts or certain kinds of

---

6 NBIC: Nanoscience, biotechnology, information technology and cognitive science. For more information and summary descriptions of some of these and the following developments and the notion of a trans-human, see the following website: http://incipientposthuman.com, e.g. http://incipientposthuman.com/physical.htm

machine parts have fewer rights than biological cell structures? Why should biological cell structure serve as a criterion for drawing a distinction between human and non-human beings? We may consider such questions far off, the progress toward these new beings overestimated, and those who believe in the posthuman challenge out of touch with the realities of science and technology. But my claim is the fantasies themselves are important. They play into and help create what I called a culture of life.

Consider now other areas of discussion. I have already pointed out that in the social sciences, "life" thinking may be implicated in those areas that have turned the individual and its search for Ego and "I"-related affirmations into topics of investigation. But a more direct incidence of a life-centred notion in the social sciences is the recent renaissance of the notion of a generation – it serves as the antithesis to the common concept of solidarity favoured by social democratic and socialist parties in the last 100 years, which is based on hopes for equality through the redistribution of resources. In contrast, generational concepts focus on individuals that are sequentially related through family ties. Hope and promise in reference to individual life also come from finance, where excess imagination – supported by the profession of financial analysts – is invested in financial scenarios as ways of enriching the self and the life course. What supports these ideas and life-course and generational thinking are institutional changes in pension schemes which move from solidarity-based principles, where income from the working population is redistributed to retirees, to personal investment schemes where one plans and pays for one's retirement benefits over the course of a lifetime. On a more conceptual and theoretical level, a return to human nature-based theories of rights and justice can be associated with life-centred ideas[7], as can Heidegger's temporal notions of

---

7 FUKUYAMA, F., 2002, *op.cit.*

human existence as "being towards death" and vitalist concepts[8] that can be linked to Bergson and Tarde.

On a popular level, the life-enhancement literature mentioned and bio-ethical controversies about the rights to genetically and technologically enrich lives and gene-lines, and the images of individuals searching for optimal experience through edgework (extreme sports etc.), suggest individuals and populations deeply involved in the appropriation of their lives and those of their offspring. For neo-Marxist thinkers, post-Fordist knowledge-based systems appropriate workers' lives rather than their labour, with work encroaching upon and difficult to distinguish from free time and coinciding with the individual's lifetime. Conflicts over the "appropriation of life"[9] rather than over the appropriation of surplus value – between economic agents, individuals, *and* the state and non-human objects (such as viruses) – may well be what defines posthuman and postsocial environments.

The Promise-Base of the Culture of Life

The language of humanism that pervades enlightenment ideas, like the language of religion or anything that requires soul, was not predisposed to accepting posthuman developments centred on life. It was not predisposed to accept the expanding role of biological and technological augmentations of human nature; drugs that miraculously change personalities or biotechnological means that permanently modify germ lines or that prevent, however indirectly, individual death. In particular, the language of humanism is not predisposed to the orientation to a future informed by continually

---

8 LASH, S., *Empire and Vitalism*, Proceedings of the Annual Meeting of the Eastern Sociological Society, Philadelphia, 2003.
9 LASH, S., 2003, *op. cit.*

emerging and constructed "promises" (rather than important and stable values) that characterises a culture of life. In speech act theory terms, the felicity conditions, that is the conditions of success of a "request" – e.g. humanism's request for virtue and ethical behaviour and enlightenment's request for reason and rationality – are quite different from the felicity conditions of a "promise". Promises must concern future acts, they must concern things that the promise receiver really wants, and they imply that the promise giver is able to and intends to keep the promise. Thus a promise giver's competence and trust in his or her sincerity, as well as the future play a role if a promise is to be successful. Fulfilment of the promise is the task of the promise giver and speaker, not of the promise receiver. We can contrast the promise scenario with that of requests, where fulfilment is the task of the receiver of the request, the wants or desires of the receiver play no role in the matter, and the future is only implicated in a trivial way – as when I ask someone to pass the salt and the passing has to be done after the demand is uttered. Competence also lies with the receiver who must have the mental and physical equipment required to comply with the request.

A promise-based culture would seem to be more seductive than a request-based culture. First, it works with people's desires, to which it pays attention and which it stimulates. Many of the life-enhancing propositions that the biological sciences in conjunction with others put before us pertain to deep-felt desires of Western cultures – the desire to alleviate illness and disease, to augment mental capacities and bodily appearance, to escape ageing and death. These desires are deeply inscribed in Western thought; they are more than manifest in literature, sculpture, painting, music, religion, philosophy, and political thinking. Secondly, since fulfilment of the promise and the requirement of sincerity lie with the promise giver, all the promise taker needs to contribute is plausible wants. Thirdly, in the present case, the promise givers are in many cases sciences. The sciences are of course not untainted by public criticism, but when it comes to beliefs in their sincerity and in their capacity to fulfil the promises they make, we trust them more than we would trust politicians, for example. Enlightenment thinking was not only based on requests but also on promises – of equality, liberty, justice, etc. The promise giver was the state, which took on the task of securing public welfare. Yet as argued before, today's Western states are gradually withdrawing from the role of the promise giver and increasingly unable to fulfil its demands; states simply lose power in a global age. The emptying out of a social imagination I have postulated has much to do with the decline of the state. Much promising has shifted, I maintain, away from the state, while requests and demands have not shifted elsewhere. States also operate increasingly out of synchrony with the promises and delivered results of the sciences.

41

● ● ● Conclusion

The move to a culture of life implies changes in regard to the source and defining concepts (the human vs. life) of our collective imagination. The analysis I have started to develop shines the analytic light on these changes, as well as on a confluence of developments in many areas that entail the rise of a culture of life. These changes are larger and run deeper than concrete ethical questions the biological sciences raise. I have suggested that they also indicate a sharp break with the enlightenment ideals of human reason and the perfectibility of society, which they replace by the idea of the perfectibility of life.

LIFE SCIENCES, THE IMAGE OF MAN AND PROGRESS

Evandro Agazzi

Evandro Agazzi is president of the International Academy of Philosophy of Science (Brussels), honorary president of the International Federation of the Philosophical Societies (FISP), honorary president of the International Institute of Philosophy (Paris), and of several other academies and learned institutions in different countries.

Within the concept of progress we can distinguish three components: the idea of change, the idea of a direction, and the idea that this change brings us closer to the attainment of one or more values. For progress to occur what is needed is a change "for the better". All this goes to show that the application of this concept is relative, i.e. it is dependent on the "good" (or the value) by reference to which the change is evaluated. This does not pose problems when the "progress" in question relates to a success of happenings taking place within a particular field for which explicit, precise evaluation criteria exist (as, for example, when we talk of progress in chemistry, or language learning). It becomes problematic, on the other hand, when the concept of progress serves to express the belief that historical events take place in a growing line of perfection. In this second meaning we are no longer talking of the progress *of* something, but we envisage progress itself, viewing it as a positive statement on our past history and as a sort of prophecy for the future.

● ● ● The general Concept of Progress

This second meaning of the concept of progress expresses an optimistic conception of the course of human history. This has become a fundamental characteristic of our "modern" mentality, which has been profoundly affected by the birth of the *natural sciences*. This explains why the idea of progress as a historical category, although prefigured in the reflections of 17th century writers (Bacon, Descartes, etc.) and of the Enlightenment philosophers (such as Voltaire, Turgot and Condorcet), dominates the 19th century vision of history, both in the version that inspired the "Romantic" philosophies of the idealists and that which welcomed the opposing doctrines of the positivists who explicitly represented science as a driver of progress. The crisis of the ideology of progress began only with the tragic experiences of the Second World War.

● ● The Evaluation of Progress and the Image of Man

There are many reasons for this crisis. We will analyse just one of them. We have already noted that the application of the concept of progress is relative to the value used as a reference point in assessing a particular change. This relativity is in danger of rendering inapplicable the concept of progress taken in its general sense. This is because the reference value needs to be something "global" and very vague, like the "good of man" or "happiness" and, at the same time, definable in a manner that transcends purely subjective personal preferences. It has frequently been thought possible to define this reference value by observing the needs, aspirations or demands inscribed in *human nature*, i.e. which present themselves as consequences of a particular vision or *image of man*, which is deemed to be solid and pertinent, and capable as such of expressing *human identity*. However, it is not difficult to observe that the crisis of the idea of progress that we have just spoken of is precisely the consequence of the disappearance of such a reference image. This reflects an "identity crisis" of modern man, frequently underscored in philosophy, the arts and literature. It is also interesting to see to what extent science in general, and life sciences in particular, have contributed to this crisis.

••• Reasons for contemporary Man's Identity Crisis

We often talk of the *loss of identity* suffered by human beings today. This phrase expresses a loss of direction, a profound crisis that has affected reference frameworks, i.e. constellations of values within which men and women have been accustomed to situate themselves and which give meaning to their existence. All this is true, but the even more radical sense of this loss of identity consists in the difficulty of answering the question: "who am I?" or, more generally: "what is man?". This is an eternal question, already alluded to in the imperative "know thyself" in which classical antiquity placed the essence of wisdom. The paradox of this invitation consists in the fact that knowledge of oneself is, *prima facie,* an easy task, in the absence of any division between the knowing subject and the object of this knowledge. This facility becomes difficult, however, if we rightly consider knowledge as an operation that *demands* an otherness, a distancing between the subject and the object. From this point on, the "objectivation" of the person presents itself as an enterprise of despair, and we feel ourselves pressed to look for indirect "resources" (that is "sources of knowledge") to achieve this. The inevitable consequence is that the response to the fundamental question will depend on the nature of the resources used, and that human identities will be reconstructed according to the characteristics that such resources are able to handle, or which such sources of knowledge are able to offer us. For centuries, the West found such resources in philosophy and religion, and sources of knowledge in metaphysics and revelation. In the modern age, a new means of knowledge has appeared on the horizon: natural science, no longer using either metaphysics or revelation. The conceptual space in which it encloses everything that it can know is limited to matter and movement (the latter being governed by "forces" which determine it by acting "from outside" on material bodies).

The dualist Vision of Man

The most attentive spirits were quick to perceive the risk implied by the uncontrolled dilatation of this new source of knowledge, and were at once concerned to "save man". The solution they adopted was a "dualist" conception of the human being, inaugurated by Descartes, but accepted in different forms by various thinkers down to Kant. This is the famous doctrine of the "two substances", according to which the human being is the result of the union of two distinct, autonomous and independent realities, the material body and the spirit. The advantage of this split appeared to consist in the ability to adopt at once a totally "modern" attitude towards the new natural sciences, whilst safeguarding the spiritual values that the metaphysical and religious traditions included in the image of man. Simply, man's *body* was viewed as belonging in the field of the natural sciences, but with the mental reservation that, in any event, the authentic man remained intact within the citadel of his subjectivity, to which natural science was without access and which remained in the sphere of competence of metaphysics and theology. Unfortunately, this reasoning led equally to the legitimisation of the inverse viewpoint: no illumination or principle from the domain of the spiritual, the metaphysical or the transcendent was considered necessary for interpreting and explaining the structure and functioning of the human body. The spiritual soul became a useless voice (a superfluous hypothesis) in explaining the functioning of the human body, including a large number of functions which can scarcely be described as mere physiological manifestations. This situation was made possible by the formidable power of representation of the machine, a typical product of the new science.

The Machine as a Model of Intelligibility

Whereas traditional technical tools were in general the fruit of more or less fortuitous "discovery", improved and refined over time by the accumulation of practical experience, *machinery* has the characteristic of being *invented*, i.e. of being first designed, and then produced. In other words, we know how it will be structured, how it will work and why it will work in a particular way "before" we begin building it. All this comes down to saying that there is no "mystery" in a machine. It is easy then to understand how the machine takes on an "epistemic" role. Not only is it a tool which is infinitely more effective than any natural products or processes, but it is also a formidable "conceptual model": once we have been able to interpret any *natural* system as a more or less complex machine, structured and operating according to the "laws" of a certain science that we have mastered, we will have the impression that we have "understood" and "explained" it in its entirety.

There is no way of denying the fascination of such a conceptual model. Its application to man became irresistible with the imposition of Cartesian dualism. Descartes himself has left us one of the first examples of the theory of *man-machine*, inaugurating an uninterrupted serious of analogous visions based on the types of machines (chemical, thermodynamic, electromagnetic, cybernetic) that new sciences have enabled us to dream up.

●●● Efforts to move beyond Dualism: Idealism and Materialism

The vicissitudes we have sketched out illustrate the historical consequences of Cartesian dualism which, in its day, sought to "save mankind" from the menaces of the materialism borne on the wave of the new mechanical sciences. By separating the two substances, and reducing the "true" essence of man to the thinking substance only, philosophers believed they had barred the route to materialism. This philosophical "imagination" proved, however, unable to weaken the strength of the "phenomenological evidence"of an *ego* having both a both and a spirit. Hence the arduous problem of understanding how the two separated substances can *interact*. The impossibility of finding a solution lay in the fact that the "compromise" of two substances already signified the loss of man's identity (which implies a *unity* in which all *differences* come together). Hence the later attempts to recover this unity through the philosophies of idealism and materialism. Unfortunately, this effort took a wrong path: accepting *duality* as the starting point, people tried to regain the unity by removing or neutralising the differences, that is by trying hard to demonstrate that, of the two permitted realities (spirit and matter), only one was the true substance, the other being only a partial and particular "manner of being". As to which of the two realities was the true substance, when all was said and done, this depended on the chosen option: idealism affirmed that the real autonomous substance is Thought or Spirit, whilst materialism stated that the true substance is matter, and that spirit is no more than a particular form taken by matter when it reaches a sufficient level of organisation and complexity.

Historically speaking, idealism lost its cultural battle, precisely because the "deduction" of the multiform variety of individual concrete phenomena from pure rational arguments proved inadequate. Does this make materialism the winner? Perhaps not yet, but it has increasingly become the conceptual space of our civilisation, thanks to its ability to use in its defence the conceptual tools of modern science and technology, developed precisely, as we have seen, for studying the *res extensa*.

●●● The ontological Impact of the Techno-sciences

In this context, the "technological" dimension regains a decisive role by offering science (of which it is the concrete application) a potent argument for affirming its ontological pretensions. If we limit ourselves to considering pure science, we can indeed always affirm that it can only offer us a logically satisfying framing of *phenomena*. However, modern science does not just seek to know and understand the world. Thanks to the technology that applies its knowledge, it is able to *modify* the world and even "create" new realities. The world of technology is an enormous system of "substances" that "did not exist" before mankind produced them, and it would be simply ridiculous to state that these technological artefacts are pure phenomena. It would appear legitimate then to state that the sole true reality is that which science studies, because it is science that allows us to intervene and considerably enlarge the wealth and frontiers of this reality. But there is more to it than this: technology enables us to "construct" material and "concrete" machines that behave like the natural systems we wish to study. These machines are, so to speak, "ontological" models and their efficacy appears to justify the thesis that natural systems do not have a different ontological status, and that there is no need for other ontological principles in order to explain them.

Applying this strategy to the modelling of natural systems (such as human beings) that are capable of so-called "mental" activities, people are beginning to build machines that can carry out the same "functions" as human beings. This, they believe, entitles them to state that these functions do not require the existence of a non-material substance (such as the spirit) to understand and explain them. In this way, the spirit is reduced to an epiphenomenon of matter: matter, once it has achieved (through the dynamics of chance and necessity) the necessary level of complexity, becomes capable of "thinking".

● ● ● The interior Experience

If human beings find in their experience of themselves reasons not to accept their being reduced to manifestations of the universal Spirit, this same experience prevents them *a fortiori* from identifying themselves with "external" nature. The fact is that human beings are capable of "interiorising" the external world whilst not ontologically identifying with it. This "internal presence" of the world constitutes the grouping of *representations*, and the faculty of producing representations is indicated in philosophy by the term "intentionality". Many animals are already endowed with this capacity, but human beings possess it in a much larger proportion, being able to represent also abstract objects, such as concepts, pure possibilities, purposes, duties, in respect of which they express "judgements", that is an adherence, an acceptance or a refusal that "depend on themselves". The fact is that the machines which have been created to "imitate" human thought, whilst able to carry out many "operations" analogous to those of thought, remain ontologically distinct from human beings in that they lack these endowments of intentionality and judgement.

Another order of "phenomenological self-evidences" that human beings find within themselves is that of "moral" self-evidence: every human being is capable of "moral judgement" according to which they have a "duty" to do good and avoid evil. Such concepts of duty-to-be and duty-to-do are totally inapplicable to the world of nature and, therefore, do not figure among the concepts and principles of science and technology. Every system, both natural and artificial, behaves according to the "constitutive rules" that determine its structure and its operation, and to which it is largely bound. However, any interpretation of man that reduces him to a machine (albeit highly complex, through the conjunction of different machines) profoundly troubles his identity by clearing him of his moral dimension. If man is no more than a "result" of certain "mechanisms", his "actions" are reduced to pure "behaviours", completely conditioned and for which one cannot ask whether or not they are in conformity with what they "ought to be".

This means, at the same time, that a pure and simple techno-scientific reading of man frustrates him of this essential aspect of his identity, that is constituted by the awareness of his "freedom", without which the very meaning of morality disappears, as do the sense of any goal in life, the purpose of responsibilities, and any form of creativity.

The techno-scientific Image of Man and the Idea of Progress

We can now ask ourselves whether this image of himself that man has drawn from the new wellspring of techno-science is able to offer him reference frameworks with which to evaluate progress. Here we are immediately confronted with a paradox. At the same time as progressively reducing mankind to the status of just another natural being, modern culture has celebrated the *dignity* of man and solemnly proclaimed a growing number of "human rights", which today form a sort of common basis of our common moral consciousness, including at the level of political morality. How can we reconcile this moral attitude with the idea that man is a machine? What sense is there in saying that one should respect a machine? Is not the rational attitude towards a machine precisely that of "exploiting" it to the full. Here we have another aspect of this loss of identity brought about by a purely techno-scientific vision of man: it cuts at the roots of the "reasons" for affirming man's dignity and his fundamental rights.

In order to understand the question it is useful to consider the favourable position of the traditional "metaphysical" (and theological) type vision which saw an "ultimate purpose" in nature, in the double meaning that every natural being had in its own essence the pattern of its "individual" purpose, and that these various purposes were harmonised in a "global" finality. Man, too, was included in this perspective: he had his "own" finality (derived from the sum of "all" his essential characteristics), and it was his task to act within the "moral order" (which, in turn, co-penetrated with the natural order).

In techno-science, on the other hand, is there is no precise finality because its progress "depends" on already acquired knowledge and methods, it advances under the "impulsion" of cognitive situations and of problems that are "behind" it. And even if we wished to find in the human will the final purpose behind techno-science, this will would be unable to foresee a "global finality" on the basis of pure techno-scientific criteria.

The Recovery of objective Interiority

We have now begun to glimpse the solution to our problem: man can recover his lost identity solely by finding again the path to his "inner self". It is in the sphere of this interior being, this *intériorité*, that can lie the self-evidence of the intentionality which, in its simplest forms, is also shared by a large part of the animal world but which, in its highest forms (thought, conscience, self-consciousness, verbal communication, development of symbols, capacity of abstract thought and sense of the possible, moral conscience, freedom of self-determination, sense of transcendence), distinguishes and separates man from the rest of the living world and, *a fortiori*, from inanimate objects, including the products of his technology, however complex and marvellous they may be. This sphere of the inner self is far from reducible purely and simply to a sort of elementary, unfathomable, obscure and inexpressible *point*. Rather, it constitutes a veritable *inner world* which can be analysed, deepened and explored, using "methodologies" that mankind has tried to define and "practise" in its different cultures, including western culture, before imperialism and techno-science led people to abandon such concerns.

One reason for this abandonment is that interiority has been confused with pure and simple "subjectivity", and that modern thought, which has discovered, on the one hand, the scope of subjectivity, has considered it, on the other, as the diaphragm "separating" us from the real world, as something "preventing" us from knowing reality. In our efforts to gain "objectivity" (considered as the opposite of subjectivity), only "science" has appeared capable of attaining a worthwhile result. As a consequence, only scientific criteria and methods have been considered as being available to achieve this endeavour.

Against this we need to recognise that the interior world is not identifiable *en bloc* with the sphere of subjectivity which encloses us like an inescapable prison. We need to re-conquer the idea of an *objective interiority*, as a reality that all men have "experienced" (even if not tangibly), and which they can speak of amongst themselves, and which they can teach other people to analyse.

The Contribution of Life Sciences

If the interior world, which is not patent of correct interpretation using scientific categories, needs to be able to become part of the "image of man", we then need to reinstate the "sources of knowledge", such as literature, poetry, the arts, philosophy and religion, that concern themselves with it. This image will, however, lack unity if science's contribution remains that of the man-machine which, as we have seen, is a direct outcome of dualism. The situation could improve considerably if life sciences were to contribute to modifying this perspective. This they could do by becoming aware of their "epistemological autonomy" vis-à-vis the physical sciences. The fact is that "living matter" is a pure abstraction. What exists are living individuals, organisms, each of which constitutes an organised unit, a system of functional relationships, consisting of a large number of components endowed with specific properties but in turn possessing properties which are novel and different from those of its components. The concepts of finality, of development towards the realisation of an internal design, of spontaneity and of uncertainty in comparison with external stimuli, have always been part of the conceptual framework of the philosophical and scientific considerations which concern living people. The cultural imperialism exercised by physics from the 18th century onwards alone has led the life sciences to abandon these concepts and to *reduce* them to mechanistic frameworks by means of forced interpretations. Today, biology is in the process of becoming once again the driving science, but is in danger of allowing itself to be once again dominated by the conceptual paradigms of physics. Concepts of complexity and emergency, of systemic organisation, which ought to serve to express the features of novelty and purposeful organisation which characterise the living world, are frequently voided of their conceptual force by reducing them to paradigms such as "determinist chaos", "order by fluctuation" and non-linearity which, ultimately, reduce the richness of the world to the blind game of certain elementary mechanisms subject to chance. This tendency is strengthened by the strictly determinist framework which emerges once the specific components of living being have been defined. What we are talking about in particular is the theory that all the features of a living being are inscribed in its genome (for example, when we talk of the "aggressiveness gene", the "altruism gene", or the "artistic creativity" gene, etc.). It is clear that, if ever this *genetic determinism* enters into the image of man that science is helping to forge, we will have perfected, instead of rectifying, the model of the man-machine which constitutes a serious obstacle to the reconstruction of a human identity.

WHAT PHILOSOPHIES ON PROGRESS FOR THE THIRD MILLENNIUM?

Gilbert Hottois

Gilbert Hottois is a professor of philosophy at the Free University of Brussels (ULB) and a fellow of the ULB's Interdisciplinary Research Centre on Bioethics (CRIB). He is also a member of several national and international scientific societies and ethics committees. His published works in French include *For a metaphilosophy of language, From common sense to the communication society,* and *The bioethical paradigm.*

Ever since the concept of progress appeared in association with science and modern technology, it has constantly occupied a highly significant and problematic position in philosophical thinking and in the collective consciousness. From the rise of modernism in the 17th century to the post-modernism of the beginning of the third millennium, we present the main stages and principal facets of this critical reflection by reference to a handful of representative philosophers: F. Bacon and R. Descartes; I. Kant, G.W.F. Hegel and K. Marx; H. Jonas; K.-O. Apel and J. Habermas; F. Fukuyama and H.T. Engelhardt.

●●● Francis Bacon and the original religious Framework behind the Idea of techno-scientific Progress

The work of Francis Bacon (1561-1626) presents the first systematic occurrence of the concept of "future generations" in direct relationship with what we call techno-scientific R&D (Research and Development) today. Bacon himself declares at the start of his *Novum Organum (NO)* that he is writing for the "benefit of present and future generations". Baconian science is an operatory science that experiments, transforms and produces. As such it is opposed to scholastic science which is purely verbal, speculative and passive. His new science is at once knowledge and power: once we know the effective causes of a phenomenon, we can produce it, prevent it or intervene to modify it. It is also a knowledge that develops progressively through a collective organisation of research. This techno-scientific collaboration needs to continue from one generation to the next. Progress is the bearer of a very great hope: that of making man into a sort of god. And it is man himself who is responsible for this hope, as it is "man who is a god for man".

Below are extracts from two of Bacon's best-known works, the *Novum Organon* (1620) and *New Atlantis* (1627) (*NA*) – which clearly set out the issues at stake.

"Man is a god to man, not only in regard to aid and benefit, but also by a comparison of condition. And this difference comes not from soil, not from climate, not from race, but from the arts. (…) Now the empire of man over things depends wholly on the arts and sciences. For we cannot command nature except by obeying her[1]."

"The end of our foundation[2] is the knowledge of causes, and secret motions of things; and the enlarging of the bounds of human empire, to the effecting of all things possible[3]." Running through the impressive list of the discoveries/inventions achieved by the searchers for the *New Atlantis*, we irresistibly recognise our current biotechnological world. For example, they make plants "by art greater much than their nature; and their fruit greater and sweeter, and of differing taste, smell, colour, and figure, from their nature. And (…)

---

1 *NO*, aph.129.
2 The House of Solomon, a sort of technocratic government (note by the author).
3 *NA*, p.119

many of them we so order as that they become of medicinal use. These transformations also relate to animals and 'new and fecund species' are created[4]."

In the list of the surprising "natural marvels, in particular those destined for human use[5]", we find the following anticipations: "prolonging life, delaying old age; healing supposedly incurable maladies; reducing pain; transforming the temperament, stature, physical features; metamorphosis of one body into another; the creation of new species; production of new foods, etc."

Francis Bacon was probably the first to formulate what is known today as the "technician's imperative[6]", which commits us to achieve everything that is possible, and which frequently appears to suggest that "nothing is technically impossible". We gain the impression that Bacon has in mind an unlimited progress which ultimately transforms man into an omniscient and omnipotent God. In fact, his thinking lies within the Judeo-Christian framework.

The nature that man manipulates has been created by God. Even if offering numerous possibilities of transformation, it is not an amorphous and infinitely plastic matter, bereft of laws and essential forms. Bacon repeatedly tells us that "we master nature only by obeying it", that is, on condition that we understand its laws and make skilful use of them in the interest of mankind: "artificial things differ from natural ones not in their form or essence, but only by their efficiency; man, in nature, having power over nothing but motion[7]." This future, placed under the sign of progress, is not part of a cosmic temporality infinitely open to unforeseeable futures. It too is encompassed by Christian mythology: making man a god is to progressively give him back the status that was his before the Fall,

---

4 *NA*, p.122-123.
5 *NA*, p.133-134.
6 In *NA*, he indeed talks of carrying out "every possible experiment" (p. 121).
7 Cited in the Introduction to *NO* (p. 29).

when he reigned over the Garden of Eden and enjoyed Adam's knowledge and power. The purpose of science and technology becomes that of progressively redeeming man – in a progression that is at once a regression – from original sin. In his *Discourse on Method* (1637), René Descartes – the other philosopher generally viewed as an originator of modern science – faintly echoes Bacon when he states that Method will render men "like masters and owners of nature". In short, the ultimate purpose of progress is to regain the lost origin: time forms a circle within divine creation.

Nor should we forget that Bacon's utopian *New Atlantis* is not a democratic society, but a sort of techno-scientocracy governed by the House of Solomon – a council of wise sages – which decides which discoveries and inventions are to be made public. In this way, Research is made distinct from Development.

● ● ● The Enlightenment, Kant, Hegel, Marx and Fukayama: the modern Philosophy of Progress

From Bacon to the 20th century, we witness a growing secularisation of the temporal mythical-religious framework of the Judeo-Christian tradition. This secularisation coincides with a politicisation which brings with it a marginalisation of the philosophy of nature, science and technology in favour of moral, legal and political philosophy. The Enlightenment and the French Revolution play a key role in this evolution, to which Immanuel Kant gives the most complete philosophical formulation in presenting religious (Christian) ideas "within the limits of pure reason". It is no longer a question of returning to man's Edenic origins, but rather the gradual accomplishment, in the course of human history, of an ideal that exists "potentially" within imperfect human nature. This ideal is that of a universal, just and pacific society, in which reason and freedom will flourish. This is the ultimate purpose of human history and, in accomplishing it, men realise the plan of nature. At least this is the reasoned philosophical hypothesis that Kant proposes to both present and future generations. The short article written in 1784 entitled *Idea of a Universal History from a Cosmopolitan Viewpoint* is particularly explicit: "We may consider the history of the human species as a whole as the accomplishment of a secret plan of nature to produce a political constitution that is internally perfect and, to this end, also externally perfect, such constitution achieving the sole situation in which nature is able to fully develop all its dispositions in humanity[8]." Kant recognises that philosophy in this way has its own "millenarianism", reasonable belief, which should help it achieve its own and very desirable realisation: "We would be able, by our rational preparation, to hasten the advent of this instant so felicitous for our descendants[9]." Kant has in mind here essentially moral, legal and political progress. The development of science and technology and the universal knowledge they dispense are auxiliaries of which nothing revolutionary or decisive is expected.

Georg Wilhem Friedrich Hegel, moving from reasonable belief to necessary knowledge, expands to the full Kant's ideas concerning the ultimate purpose of nature and the theologico-anthropological eschatology of history, ending up in the advent of the "universal state" and of absolute knowledge.

Disputing Hegel's idealism, Karl Marx was to complete the process of secularisation by considering only the historical, material production of the classless society, permitting the fulfilment of all humanity in each individual. This, for him, is the ultimate "end of history", to be

---

8 Eigth proposition.
9 *Op. cit.*, p. 201.

realised by political revolution followed by the non-alienating management of the economy, encompassing technology (production methods). These "joyful futures" are not for a far distant future. It is the sure knowledge of their nearness in time which gives hope to humanity.

At the other end of the political philosophical spectrum, with the 20th century entering its final decade, Francis Fukuyama believed he could declare the end of history to have already been achieved in a liberal democratic society with a market economy[10]. Such a society achieves the ideas of the 1789 revolution by recognising the equality in law of all human beings. There is no longer any need to expect or hope for the emergence of a new political regime or the return to an earlier regime. Its advent and its expansion have been facilitated by scientific and technological progress during the past two centuries.

Very recently, in *Our Posthuman Future. Consequences of the Biotechnology Revolution*[11], Fukuyama, who has since become a member of the United States' Bioethical Commission[12], has realised that techno-scientific R&D, in particular in the biomedical field, is in danger of jeopardising this final state, when techno-science becomes capable of manipulating and transforming human nature, on the permanence of which the final social, legal and political state of history rests. "There cannot be an end in history", he insists, "unless there is an end in science[13]." His work seeks, in various ways, to rescue the stability of human nature threatened by technosciences[14].

---

10 FUYAMA, F., *The End of History and the Last Man*, New York, Free Press, 1992.
11 FUKUYAMA, F., *Our Posthuman Future: Consequences of the Biotechnology Revolution*, New York, Farrar, Straus and Giroux, 2002.
12 Established by President Bush.
13 FUKUYAMA, F., 2002, *op. cit.*, preface.
14 See our critical study: CHABOT, P. & HOTTOIS, G. (Eds), "La 'fin de l'histoire' excédée par la recherche techno-scientifique", *Les philosophers et la technique*, Paris, Vrin, 2003.

● ● ● Hans Jonas and the Critique of the modern Philosophy of Progress

Everything we see around us tells us that the Darwinian revolution and its evolutionary consequences have not yet affected, or are only beginning to affect political philosophy, the first reactions of which are defensive, conservative or reactionary. The implications of evolutionism are found at two levels:

(1) the biological level, leading to the "naturisation" of man: man is one living species amongst others, the non-necessary product of an immensely long evolution; and

(2) that of biotechnology moving towards the "operation(alisation)" of man: as a non-necessary product of evolution, man can change technologically the natural being that he is.

From the first level to the second, we move from a science that describes and explains to a technoscience that produces, transforms and creates. It is a new awareness of these implications that motivates Fukuyama's above-mentioned reaction, after many others, faced with the increasing ambivalence of technology and human progress during the second half of the 20th century. This takes place at two levels: (1) empirically and in relation to man as he is: harm of every kind (from pollution to weapons of mass destruction, unemployment and ecosystemic imbalances); and (2) essentially and in relation to the possibilities – procreative, genetic, prothetic, neurotechnoscientific – for the transformation of man as he currently is.

The Imperative of Responsibility. In Search of Ethics for a Technological Age, published by Hans Jonas in 1979[15], constitutes the most systematic philosophical reaction to the dangers for the future of humanity and nature from the progress of technoscientific R&D in democratic, multicultural and individualistic societies.

---

15 JONAS, H., *The Imperative of Responsibility. In Search of Ethics for a Technological Age,* University of Chicago Press, 1985.

Jonas starts by observing the new scope of action, hugely multiplied by the technical power and number of human players involved: the impact on nature and society is ever deeper and broader, the consequences extending ever further in a future that is becoming increasingly difficult to anticipate.

His other initial observation is the sense of nihilism in the air: the disappearance of universal belief in absolute values and references, based on religion or metaphysics. Today, all values and standards appear to be decided by human beings, on an individual or collective basis. This anthropocentrism is expressed in cultural and historical relativism, in the individualism and pluralism of democratic societies. In the absence of transcendent limits or interdictions, in principle there is nothing to stop anything that is possible being tried one day.

This conjunction of technological super-powerfulness and nihilist freedom will place nature and humanity in grave danger of essential disfiguration and even of destruction.

In response to this threat, according to Jonas we need an ethics that is at once solidly anchored and attentive to the future. This will provide the basis for politics governed by wise men whose dominant concern will be to protect human existence and the human essence in the long term. As a foundation for these ethics and politics, Jonas develops an ontology and a philosophy of nature which describe the evolution of life as a purposeful process, culminating in the human race. The latter is at the summit because man is the being of purposefulness, the one who, freely, intentionally, invents and carries out projects. For Jonas, natural evolution is characterised by the successive appearance of living species that are increasingly capable of intentional behaviour, as if freedom – which is also the mark of the spirit – was affirmed progressively via the evolution of living beings[16]. In other words, the

---

16 See our analysis : Hottois, G. & Pinsart, M.-G., "Le néo-finalisme dans la philosophie de Jonas", Hans Jonas. *Nature et Responsabilité*, Paris, Vrin, 1993.

value of natural-cultural man is rooted in the very act of becoming of being and of nature (in another era Jonas would have said: in the creative will of God). This is a value *in se* which imposes itself on human beings and cannot depend on their own choices and decisions. The mission of ethics and the politics of responsibility are therefore to forestall any enterprise, in particular in the field of R&D, which carries any risk, however distant, of disfiguring or destroying humanity directly or indirectly, or excessively mistreating nature – which is one of the preconditions for the existence of the human being. To be feared in particular are the manipulation of the means of human reproduction, and genetics.

For Jonas, the idea of progress and the humanism of the Enlightenment, which root meaning and value in human reason and freedom and which set out to dominate nature, have led to the techno-scientifically armed nihilism that characterises the 20th century. In order to combat these dangers, Jonas exposes himself to the question of evolution. But this remains a feigned exposure, since the implications of neo-Darwinian evolutionism (contingency and manipulability of nature; "naturing" and operation of humankind) appear even more nihilistic than those of the historicist philosophies of progress. Jonas refuses two aspects of evolutionism: One relates to the past: the contingency of evolution, rejected by the affirmation of final purpose; the other relates to the future: this is the freedom of the human race to effectively intervene in its own evolution, even if a time abyss of millions and billions of years opens up in front of it. Given this abyss-like future, it is difficult to accept Jonas' belief that the human species, as it has existed since prehistory, constitutes the ultimate end of bio-cosmic evolution, and that our sole task, hope and desire should be to preserve it.

The paradox of Jonas' thought it is that, on the one hand, it justifies the eminence of man by emphasising freedom and humankind's ability to give itself goals and to transcend given situations while, on the other hand, it simultaneously confines this freedom to the field of symbolic invention and creation. This can be understood solely against the background of religious convictions that remind us that man is made only "in the image of the Creator", that he is a simple creature and as such may be freely creative only in the field of images, symbolic representations and names. He is not permitted to intervene in the onto-theological process of creation that natural evolution expresses, and even less so in his own creation, by prolonging or adapting this process.

Our sole fundamental duty towards future generations is to leave them a world in which natural-cultural human existence defined in this way remains possible. Ours is a duty of preservation, not of progress. Philosophically, Jonas' onto-theological thought points towards a pre-modern and pre-critical past.

●●● Karl-Otto Apel and Jürgen Habermas: the Reaffirmation of the modern Philosophy of Progress

Apel and Habermas[17] propose a philosophy and a procedural ethics of the discussion, continuing modern thought, enriched by the contributions of the philosophies of language and history almost totally absent from Kantian thought. This philosophy and ethics denounce the Jonassian return to a pre-modern thought dictated by the symbolic and physical dangers associated with the modern philosophy of progress. For them such a return is tantamount to

---

17 We refer the reader principally to two texts in which K.-O. Apel defines his position in relation to Jonas: "The Problem of a Macroethic of Responsibility to the Future in the Crisis of Technological Civilization: an Attempt to Come to Terms with Hans Jonas's *Principle of Responsibility*", Man and World, 20, 1987; and HOTTOIS, G. & PINSART, M.-G. "La crise écologique en tant que problème pour l'éthique du discours", 1993, *op.cit.* For Habermas, see in particular: HABERMAS, J., *The Future of Human Nature*, Cambridge, Polity Press, 2003.

throwing out the baby with the bathwater, with risks greater than the dangers it is seeking to prevent. Irrational fundamentalism asserts the right to close the discussion and can lead only to violent conflicts between individuals and collectivities who do not share the same beliefs. Biocentrism, by denouncing modern anthropocentric excesses, reduces the concern of humanity to a concern for survival and conservation. Such a concern can certainly find a place in a fossilised, unequal and dogmatic society. Is not Jonas' real desire for non-democratic politics, given all the compromises that democracy has made with the modern idea of progress and the alleged need for urgent measures to conserve nature and human nature, both of which are in immediate peril? Apel denounces this apocalyptic fear which, if justified, could alone legitimate a non-democratic state of emergency and public safety measures, even going so far as, if necessary, the abrogation of human rights. He rejects the excesses of Jonassian technoscientophobia as well as the reduction of the modern idea of progress to technoscientific advances in the field of objectification, control and mastery. For him, giving up the modern idea and concern for universal progress is tantamount to abandoning the ideals of justice, equality and emancipation, as well as the political desire to realise them progressively for humanity as a whole, in particular by combating non-democratic regimes.

Apel and Habermas defend the possibility and necessity of rationally justifying every decision, particularly those concerning large numbers of peoples, or indeed humanity as a whole (including the generations to come who can be "represented"). The procedure to be respected is argued discussion between all interested parties (or their representatives), undertaken without constraint and as long as is needed to arrive at a consensus freely adopted by all. The presuppositions underlying this philosophy and ethic are: man as man is the being of language-based communication. All language-based communication involves a claim to truth, i.e. the adherence of all speaking subjects to what is being communicated. This universal adherence is the criterion of reason. It is in serious, well-argued, open and unconstrained discussion that this rationality of communication comes about most effectively. Apel and Habermas are well aware that the conditions of universal, unconstrained discussion are never more than very imperfectly achieved in concrete debate. Even so, they are achieved to a greater extent in developed democratic societies than in authoritarian and materially underdeveloped ones. For this reason, the philosophy of discussion contains a normative ideal which one should seek to achieve progressively. Such progress involves improving human institutions in the direction of greater equality, freedom and justice, and fewer irrational constraints from the political, economic and material organisation of society. R&D has an important role to play in this process of emancipation, providing that it does not lose sight of the ideal to be achieved: a universal society of unconstrained communication, made up of free and rational individuals. Taking the future generations as seriously (as potential talking partners) as we can today, our duty or our responsibility are to make sure not that they simply exist but that they enjoy social and material conditions more propitious to the practice of the ethics of discussion and to the forward progress of reason in history.

The philosophical tradition that Apel and Habermas promulgate has its roots in Kantian and Hegelian idealism. The concept of prehistorical evolution is almost totally absent. Human history, the only history that counts philosophically, is certainly progressing thanks also to the improvements in living conditions brought about by science and technology, but it is essentially a matter of spiritual and moral progress. There is also relational progress, in that what we are dealing with is social and political progress which concerns fundamentally the quality of symbolic exchanges between speaking subjects. The fundamental presupposition of this updated idealist philosophical tradition is that this progress can be achieved only by symbolic methods and by language (the logos, the seat of reason and freedom). Mankind is the living being endowed with speech, the symbolic animal, the *"zoon logon echon"* who, as such, has no need to perfect itself or to realise itself other than by this essential faculty, thanks to which it is also, always and already, ontologically distinct from all other living beings[18].

Wanting the progress of man as man to take place other than through language is illusory and dangerous. However, it is the technophysical means and resources of progression that certain partisans of the technosciences have in mind: genetic manipulation, neurotechnology, prothetics, etc., ending up with a technocratic organisation of society. The conviction of idealism, both traditional and contemporary, is that such undertakings cannot be emancipatory, or favour the fulfilment of freedom, reason and the spirit. On the contrary, they will alienate and disfigure mankind in the direction of animal or mechanical inhumanity. Biophysical and material evolution has ended with the advent of man. It is not for us, with our technology, to pick up, correct or prolong evolution, where it left off. There is no human *evolution*; there is only human *history* which is fundamentally symbolico-spiritual and not techno-physical.

●●● H.T. Engelhardt and symbolico-technical Post-modernity: Progress in the plural and beyond Progress

Recent neo-Darwinian conceptions of evolution do not perceive it as a unified process with a single, clear finality. Rather they describe it as contingent and unpredictable, throwing out random shoots,

---

18 With the exception of the concept of progress, this fundamental presupposition applies also to the phenomenological and hermeneutic currents of thought (Husserl, Heidegger, Gadamer, etc.) and to certain post-modern currents (e.g. Richard Rorty and "conversation philosophy").

unforeseeable. The human species, for such philosophers, is no more than an improbable form of life and intelligence, appearing through the random effects of cosmic evolution, in a totally localised space of the universe and of time, that is the planet Earth and its nature. The ancient and current pretensions of *homo* to universality and eminence within the cosmos are, for them, meaningless and narcissistic. Given that the human species has always been genetically and culturally polymorphous, there is no reason, they say, why, given the immensity of the future, this species should not undertake to intervene in its own evolution, once it has developed the techno-scientific capacities to do so. There is no reason why such interventions could not be multiple and varied, leading the human species towards transformations and a biophysical and cultural diversification that we are unable to anticipate either by reason or even in our imagination. There is no reason why such deliberate mutations and transformations should necessarily lead to an alienation or disappearance of freedom, conscience and the spirit.

H.Tristam Engelhardt arrives at such conclusions by taking seriously the context of contemporary biomedicine: that of a multicultural and technoscientifically evolutive civilisation[19]. For him, the modern dream of a single, universal progress has failed. Nor can it succeed because it carries particular moral convictions – those of substantial morality – which cannot be made universal. In fact, it is a sort of religion of reason derived from Christianity, which was first secularised by Kant and the Enlightenment, then by Hegel and German idealism, and finally, quasi-materialised by Marx, before being further revised and modified by Apel and Habermas. This current movement substitutes, for the City of God of the Christian end

---

19 For an exposé on Engelhardt, we refer the reader to the excellent MA dissertation by Quoilin, B., *Ethique et Postmodernité chez H.T. Engelhardt*, Université Libre de Bruxelles, 2002-2003; see also the collective work: Hottois, G. (Ed.), *Aux fondements d'une éthique contemporaine : H. Jonas and H.T. Engelhardt*, Paris, Vrin, 1993.

of time, the classless society or emancipated communication society at the end of history[20]. However, for Engelhardt, cultural and moral diversity and the variety of forms of human life are irreducible, other than temporarily, and by force, violence and constraint. As the history of the 20th century has shown, this danger of abuse is not absent from the ideologies of reason which have led to totalitarianism. The world is peopled with individuals and communities who do not share common values or the same hierarchies of values and standards, who are "moral strangers" to one another. "The apparent philosophical unanimity found in certain bioethical commissions is, at best, a political construction[21]." These "moral strangers" can, however, communicate, understand each other up to a certain point, and examine which rules they wish to share if they desire to form together a larger society, to trade and to co-operate. But this broad possibility of intercommunication is not the expression of reason marching towards a universal society agreeing on a substantially unique ethic, with a single concept of good and a single hierarchy of values and standards. The only possibility it offers is that of a procedural and formal meta-ethics which, for Engelhardt, constitutes the sole viable heritage of modernity. The only requirement of such an ethic is to settle conflicts of values in a non-violent manner, in accordance with agreed negotiating procedures, without forcing a consensus which, indeed, is never more than a limited and temporary consensus and can be called into question according to rules that have also been agreed. The minimum demand is for a peaceful and logical approach in which commitments freely entered into are respected.

20 "If it were truly possible to deduce such universally sharable truths from Reason, it would effectively be permitted to dream the crazy dream of the advent of a finally reconciled world in which all moral, philosophical and political disputes could be settled by enlightened recourse to rational arguments." In ENGELHARDT, H.T., *Bioéthique: Jusqu'où faut-il légiférer?*, Paris, Institut Euro 92, 1992.
21 ENGELHARDT H.T., *The Foundations of Christian Bioethics*, Amsterdam, Swets & Zeitlinger Publishers, 2000, p.22.

Alongside this formal and procedural meta-ethics, of potential universal application, but bereft of content, there exist an indefinite number of particular, individual and community philosophical conceptions, each with its own vision of the world and the future, with its own symbols, customs, knowledge and techniques. Lay, rational, progressive philosophy is one such vision. But there are many others, some of ancient tradition such as the orthodox Christianity to which Engelhardt subscribes. None of these has the capacity nor, in particular, the right, to impose itself by force on all others. In a certain way we have to learn to live and think on two levels: the level of formal and procedural meta-ethics which invites us to relativise our absolutes and transcendences and to avoid wars of religion (or of ideology), and the level of the particular philosophical community to which a person has freely committed him- or herself. Engelhardt does not encourage a post-modern attitude for its own sake. The post-modernist ethic consists of playing as far as possible on the diversity of the world, without any deep or lasting commitment, like an aesthete or dilettante, far from any transcendence and any sense other than for the interest of the passing moment. What Engelhardt is proposing is a strong communitarism, but one which wishes to and can remain peaceful in so far as communities and individuals cohabit, without being dissolved into it – in the post-modern space of tolerance and freedom guaranteed by the common adhesion to a formal, procedural ethics, or at least to its basic principle of not resorting to force and constraint in order to impose particular preferences and values or to settle conflicts.

Engelhardt is perfectly aware that our civilisation in not just multicultural, but also technoscientific. Discoveries and inventions are constantly posing questions to both traditional and rationalist-modern communities. These questions are expressed in the form of bioethical problems, leading us to accept or reject certain forms of biomedical and biotechnological research and applications. Engelhardt does not see any foundation of universalisable reason which could justify the absolute and universal ban on certain types of research and certain applications once individuals or communities of individuals voluntarily decide to undertake them. Taking the long-term view, as the technosciences themselves invite us to do, the space in which these forms of life cohabit could take on very different colours from those of the traditional multiculturalism that we know today. Nor is there anything to prevent a situation in which, in the long term, one form of life develops knowledge-power, and hence an intelligent physical force in a category all of its own. Engelhardt does not appear concerned by the fact that such a force might no longer recognise the meta-ethic of respect of the other, just like our own lack of consideration for the lower forms of life, both animal and vegetable, surrounding us.

We describe as *technosymbolic post-modernity* this future that we are already part way towards because it proceeds from interactions between symbolisations (traditional, modern or post-modern) and material technologies (traditional *low-tech* ones or hypersophisticated and hyper-powerful *high-tech* ones) that are increasingly capable of objectivising and operating on nature as much as the subjects of symbolisation and technoscientific invention, i.e. human beings.

The following extract from *Foundations of Bioethics*[22] illustrates this vision:

"If we take the long term seriously, major changes will be inevitable if we remain a free species advancing technologically. Human beings, Homo sapiens, have been on this earth for less than half a million years and do not share a moral vision of humanity. If we have descendants who survive into the next millions of years (a short period in geological time), it is highly likely that, sooner or later,

---

22 ENGELHARDT, H.T., *The Foundations of Bioethics*, Oxford University Press, 2nd edition, 1996, p. 417 (1st edition, 1986).

certain people will decide to remodel themselves in order to live better in the transformed terrestrial environment, and perhaps in those of other planets. Some of them will be simply attracted by the diverse possibilities for the improvement and reshaping of human nature. For others, certain possible transformations will be morally incorrect. But what is there to stop everyone from undertaking genetic interventions that will, in the long term, be both safe and readily available, in the absence of centuries-old moral foundations banning them? In the long term (…) there is no reason to presume that a single species will proceed from our own. There could be as many species as there will be opportunities inviting us to substantially remodel human nature, for this environment or for others, just as there will be reasons to refuse to take part."

Engelhardt has criticised the desire to place an absolute and universal ban on any intervention involving the human genome and, in particular, the germ line that commits future generations. Such a ban is not defensible on the sole basis of age-old procedural ethics. It postulates a religious or metaphysical foundation rendering sacred or ontologising the genome. For many people, however, the human genome is nothing but the unpredictable, contingent product of terrestrial biological evolution, not animated by any finality of which the human genome expresses the ultimate and culminating point. As a product selected by an incalculable number of random mutations, the genome is more or less well adapted to the survival of a certain form of life in a terrestrial environment. But one could well image the possibility of provoking mutations that could either improve this natural adaptation or render the genome better adapted to new environments, for example, hypertechnical ones, very far from the natural/traditional environment in which the genome has prospered for hundreds of thousands of years. "We have", Engelhardt writes, "a duty of mistrust towards the setting up of any international guardian of the genome. Not only is this immoral, but very great danger lurks in the desire to impose on everybody one particular conception of morality or a specific philosophical ideology as a substitute for religious faith[23]." Just as there is no unanimity on any absolute ban, so there is no agreement on the objectives (therapeutic, palliative, preventive, eugenic, aesthetic) of any manipulations. They are dependent on individual philosophies, each having a substantial view of what man is or ought to be, as well as of his good. This situation of broad unknowing calls for the greatest circumspection. As we are dealing with future peoples about whose conditions of existence we neither know nor can anticipate anything, we must respect is a number of minimal requirements. These are: (1) not doing anything that is likely to end up reducing the capacity of choice of future generations; and (2) not doing anything that could potentially lead to additional suffering. But *a contrario*, neither (1) nor (2) means not intervening, and *"leaving everything to nature, chance and the individual reproductive choices of individuals and communities"*. Chance (genetic or otherwise) is not freedom, nor does it necessarily help protect or promote freedom (or reason). To think otherwise amounts to assimilating chance either to the expression of an unfathomable Providence (natural or divine), or to the freedom of indifference of the irrational gratuitous act, on a totally different scale from the capacity to choose in an informed, deliberate and responsible fashion.

---

23 AGIUS, E. ET BUSUTTIL, S, "Human Nature Genetically Re-Engineered: Moral Responsibilities to Future Generations", *Germ-Line Interventions and our Responsibilities to Future Generations*, Dordrecht, Kluwer, 1998.

The Challenge and Limitations of Reductionism in Life Sciences Research

Reductionism is the name given to attempts to explain complicated phenomena in terms of simpler concepts, and it has always had an integral and powerful role to play in scientific exploration. Yet from the more "holistic" perspective appropriate to the social sciences and humanities, it can make biologists look too narrow-minded; undervaluing the impact of external forces on how organisms develop and changes occur. Some people argue that we need to look beyond the individual organism and consider the natural context of whatever we study. A more comprehensive hypothesis would argue that organisms, species and communities arise through a process of progressive development from simple to complex structures, moulded by and interacting with their environment.

Human behaviour is always embedded in a social environment that already exists, and most of it is strongly influenced by the way people learn and the culture they live in.

Some popular but extreme versions of "genetic reductionism" have led people to believe that a particular gene (the gay gene, the criminal gene, the risk-taking gene, etc.) may be directly responsible for a particular form of human behaviour. But even behaviour in which genes play a strong part is almost certain to derive from interaction between the products of many different genes. At the very most, all we can say is that particular genes may have some effect on individuals' predisposition to act and learn in a certain way.

Should biologists take responsibility for the birth of a new form of secular superstition known as "genetic essentialism"? And how does such a creed relate to the sense of violation and harm that many people feel in regard to all projects for genetic modification?

REDUCTIONISM:
METHODOLOGICAL AND IDEOLOGICAL

Fraser Watts

Fraser Watts is a professor at the Faculty of Divinity, Cambridge University (UK). Trained as a clinical psychologist and psychiatrist, he has variously worked as head of clinical psychology at King's College Hospital (UK) and as a senior scientist at the Medical Research Council's Applied Psychology Unit (UK). Following his ordination as a priest, Revd Dr Watts took up a lectureship in divinity at Cambridge University. During his career, he has published books and articles on a wide range of psychological questions including cognition and emotion and information processing in anxiety.

The public is ambivalent about science. On the one hand, quality of life has been improved out of all recognition by advances in engineering and medicine. Nevertheless, there has also been a growing scepticism about science, a recognition that science cannot solve all problems, and that sometimes it creates problems as well as solving them. There is also a growing suspicion that some of the claims of science are over-blown and cannot be trusted. Faced with public ambivalence about science, there is a tendency for the scientific community to make ever stronger and more dogmatic claims. However, that exacerbates the problem. The more the public senses that exaggerated claims are being made for science, the more sceptical about science it becomes. I believe that the best way of rebuilding public confidence in science is to be more humble and circumspect about scientific claims. In particular, I will suggest that reductionist tendencies are a major source of suspicion about science. An approach to science that was open-minded and less dogmatic would get a better public reception. It would also be helpful for science to become more pluralistic, in the sense of recognising the range of possible scientific approaches rather than insisting on the correctness of any single approach. After an initial examination of the nature of *reductionism*, reductionist ideology will be explored in three current areas of science: evolutionary biology, neuroscience, and artificial intelligence.

● ● ● Reductionism

Science as it is currently practised is often excessively narrow in its approach, and is often constrained by dogmatic preconceptions about what counts as a satisfactory scientific explanation. These preconceptions generally favour reductionist explanations, and are opposed to non-reductionist lines of scientific inquiry. It is a serious matter when science becomes dogmatic and prejudiced, rather than following where open-minded empirical inquiry may lead. Scientific practice is then no longer faithful to the genuinely scientific spirit. Also, if the public senses that science is too much influenced by dogmatic preconceptions, it only serves to strengthen scepticism about science.

Reductionism is the attempt to explain higher-level phenomena in terms of lower-level ones. Thus, for example, psychological phenomena are explained in terms of biology, and biological processes are explained in terms of physics. The attempt to explain phenomena in terms of lower-level processes is central to the scientific enterprise, and I see no reason to object to it. However, there are other ideological aspects of reductionism that are more open to objection, because they rest on dogmatic assumptions rather than open-minded empirical inquiry.

The key distinction I wish to make here is between methodological and ideological reductionism. Methodological reductionism is the attempt to develop reductionist explanations through empirical research, in as far as that proves possible. Ideological reductionism asserts that reductionist explanations must be possible, and that no other explanations are required or admissible. I am a supporter of methodological reductionism, but an opponent of ideological reductionism.

A key ideological assumption of reductionism is that the only kind of explanation that is acceptable in science is one that explains higher-level phenomena in terms of lower-level processes. However, it seems likely that there are top-down causal processes as well as bottom-up causal processes. Whether that is so is a matter to be settled by research in particular cases, not something to be excluded in advance. Another important issue concerns the degree of confidence that is attached to bottom-up explanations. A dogmatic assumption is often made that a complete bottom-up explanation must be possible. However, that is very problematic. In fact, science has probably achieved complete explanations only very rarely, and to assume that complete explanations of a bottom-up kind are in principle available in all cases is to go far beyond the evidence.

Nevertheless, on the back of this assumption that high-level phenomena are completely explicable in bottom-up terms, it is often assumed that the higher-level phenomena have no causal influence or efficacy. We are told that it may be tempting to imagine that they have causal efficacy, but it is an illusion and they are really only "epiphenomena". Thomas Huxley's memorable example of an epiphenomenon is the whistle of a steam engine. It is tempting to imagine that it is doing causal work in moving the engine along, whereas in fact it is irrelevant to it.

Even beyond the claim that high-level phenomena have no efficacy, doubts are sometimes expressed about whether they are "real". However, it is hard to know even what is meant by that suggestion. As we will see later, there are some particular aspects of human functioning that have been particularly vulnerable to the charge of not being quite "real".

To make things more concrete and specific, it will be helpful to look at three sets of reductionist ideas about human nature that can be found in contemporary science. They arise from evolutionary biology, neuroscience, and artificial intelligence.

● ● ● Evolutionary Biology

Concerning evolutionary biology, let me first make it clear that I share the assumption that species have evolved from one another. The complete evolution of a new species is something that presumably happens only rather slowly, and to the best of my knowledge we have no fully observed cases of it happening. However, we certainly have well-documented cases of a change in environmental circumstances, operating through natural selection, producing variation in an existing species. Perhaps the best-known example is when industrial pollution gave black moths a selection advantage over white moths.

Within evolutionary theory, there is a significant debate about the nature of evolution. Natural selection is clearly a crucial mechanism. However, the debate is between those who assume that natural selection explains everything and those who sense that there are aspects of the evolutionary record that are not readily explicable in terms of natural selection. Stephen Jay Gould has become the best-known protagonist of the latter position. For example, evolution seems to proceed with "punctuated equilibrium", i.e. in fits and starts. To handle that we probably need to include a more adequate theory of how environmental change has impacted on the evolutionary process. I suspect that the evolutionary theory we have at present is rather crude and simplistic, and that we will eventually need a more subtle and complex evolutionary theory that includes more than natural selection. Evolutionary reductionism is not open to that possibility, and wants to explain everything solely in terms of natural selection. There is also a debate about directionality in evolution. Many evolutionary theorists have become squeamish about the assumption that human beings represent a pinnacle of evolution; that seems to them to represent a departure from scientific objectivity. However, it is surely undeniable that human beings have not only adapted well to their environment, but have gained a significant degree of control over the environment in a way that probably no previous species has. Indeed, that degree of environmental control is so marked that it is not clear that human beings are still subject to the processes of natural selection as we know them.

Also, it seems clear that there is some kind of direction in evolution. You could not invert the fossil record and claim it had no implications for our understanding of evolution. Having said that, there is no unanimity about how best to characterise the direction of evolution. Some have suggested that it is a progression from simple to complex species. However, I tend to side with those who see a progressively more sophisticated capacity for processing information about the environment as the key directional change. It is not obvious why complexity should have an advantage in natural selection; it is much clearer why advances in information-processing capacity should do so.

Another key debate in evolutionary theorising is about chance: whether there was any inevitability about evolution moving in the direction it did. Here again, Stephen Jay Gould is a key protagonist, and on the side of those who argue that evolution is a purely chance process. Reductionist ideology is always attracted to explanations in terms of chance. However, it is arguable that it was inevitable that evolution should move towards species with an increasing capacity for information processing. That is not to claim that evolution was bound to take exactly the course it did, but only that species such as ourselves, with a sophisticated capacity for information processing, were bound to emerge sooner or later, and that when they did they would have a massive natural selection advantage. Simon Conway-Morris has emerged as a powerful exponent of the view that there was a kind of evolutionary inevitability about human beings, and he uses the phenomenon of "convergence" in evolution, the repetition of the same evolutionary developments in independent contexts, to argue for the non-randomness of evolution.

If directionality is accepted, the next question is why it happened. There are those such as Stuart Kauffman who suggest that there is something inherent in the evolutionary process that leads in the direction of complexity. That is, for the moment, a highly speculative idea. However, it should not be ruled out simply because it sits uneasily with reductionist preconceptions.

Reductionist ideology in evolutionary theorising takes the form of the assertion that all higher aspects of human nature are purely the product of evolution, or just "survival machines for our genes". Sociobiology asserted that for social behaviour, and evolutionary psychology is now asserting it for the human mind. Morality has been a key battleground of these debates.

Again, I assume that there is no doubt that morality has evolved, and that it is fruitful to place the human capacity for moral functioning in evolutionary context. I also assume that there are precursors to be found in other species. However, it seems that moral functioning takes a significantly new turn in human beings. That is linked to the human capacity for reflective self-consciousness, which most scientists would now regard as the key distinguishing feature of human beings. Because of self-consciousness, human beings have a capacity for good and evil that is more knowing and deliberate than anything that other species are capable of. Humans also seem capable of an ideological commitment to altruism that extends beyond the reciprocal altruism and kin-specific altruism of which there are many examples in other species.

There has been a regrettable tendency in evolutionary theorising about morality to play down these distinctive features of human moral functioning, and to suggest that human morality is nothing but another example of the kinds of altruism that can be found in other species. This illustrates the distinction between narrow, reductionist, evolutionary theorising that asserts that even the highest aspects of human nature are nothing but a product of evolution, and more open-minded evolutionary theorising that places human functioning in evolutionary context, but is willing to admit distinctive human features, and willing to admit a broad theoretical approach that allows for processes other than natural selection.

⬤ ⬤ ⬤ Neuroscience and human Consciousness

The next example of reductionist ideology we will consider is about the explanation of human consciousness in terms of brain processes. That is currently the focus of a huge amount of interdisciplinary scientific interest. Consciousness is one of the last remaining mysteries for science to solve. So far, it is proving rather recalcitrant to scientific investigation. However, I assume that in due course we will have a satisfactory theory of how human consciousness is grounded in the physical brain. It seems inevitable that the brain is involved somehow or other in people becoming conscious of things around them. In due course we will no doubt make headway with understanding exactly how that happens.

Meanwhile, we should be careful not to make assumptions about the nature of the relationship between brain and consciousness. There is a danger of reductionist ideology announcing in advance of any actual scientific discoveries that consciousness will be "completely" explicable in terms of brain processes. This is one of the key areas where the doctrine of a "complete explanation" is riding strong, despite the enormous problems science has had in producing complete explanations of anything so complex as consciousness. Though it seems inevitable that the brain will play a key part in the explanation of consciousness, it is premature to rule out other factors. The philosophical problem of how best to conceptualise mind-brain relationships is enormously complex. However, one helpful approach is to assume a "dual-aspect monism" in which there is a single reality that has two aspects, one physical and one mental. On this view, the integration between the two is so total that it makes no sense to talk about one causing the other. Rather, we should talk about changes in an integrated, two-faceted brain-mind system.

However, many would wish to take an alternative view in which brain and mind are seen as being sufficiently distinct for one to have causal effects on the other. If so, we need to allow for the possibility of "top-down" (i.e. mind-brain) causal factors as well as bottom-up (brain-mind) factors. We have abundant evidence, for example from the effects of head injuries, of how traumatic damage to the brain can affect mental processes. There is thus no doubt about brain-mind causal effects. However, the possibility should be left open that there are also mind-brain effects as well. The key data here is about plasticity in the structure of the brain, and how the relative size of various areas of the brain is affected by usage and experience. Here again, I am not so much arguing that mind-brain effects have yet been scientifically established, but rather just arguing that they should not be ruled out in advance of adequate scientific investigation, simply on the basis of reductionist ideology.

On the back of the assumption that we will get a complete neural explanation of consciousness, some, such as Francis Crick and Dan Dennett, have already started to draw philosophical conclusions, even before the scientific research has been accomplished. Dan Dennett makes a move that is quite common in reductionist ideology, which is to try to simplify the scientific task by minimising what needs to be explained. Consciousness is a multi-faceted phenomenon. The most basic aspect of it, one shared with other species, and the one most amenable to scientific explanation, is the basic registering of information about the environment. On top of that, there is also (a) the distinctively human reflective self-consciousness to which I have already referred, and (b) the actual experience of something, experiencing "qualia" as philosophers call it. Dennett tries to simplify the task of explaining consciousness scientifically by dismissing qualia as a mirage that needs no explanation.

This typically reductionist move is unscientific, and has no basis whatsoever in research. In fact, there are growing scientific reasons for accepting the reality of qualia. Two lines of research are particularly relevant. One is Stephen Kosslyn's research on mental imagery, which demonstrated imagery phenomena that are inexplicable without the assumption that people actually have a visual image. There is also the intriguing phenomenon of synaesthesia, in which sensory stimulation in one sense modality produces an experience in another modality. That also is incompatible with the reductionist dismissal of the reality of mental experience.

There are many different forms of synaesthesia, but the most common involves non-visual stimuli being experienced as colours. There are good demonstrations of the reliability of the phenomenon, including tasks in which people with synaesthesia are able to perform much better than people without it. For example, they can effortlessly pick out one particular shape from a jumble of different shapes because they see the shapes in different colours. *Reductionism* wants to dismiss the reality of qualia, but open-minded scientific investigation is establishing their reality.

Francis Crick is more accepting of the reality of consciousness. However, on the basis of assuming that we will have a complete neural explanation of it, he makes another characteristically reductionist move. He assumes that consciousness, and indeed personality and "soul" as well, are just products of our nervous system. That enables him to draw the astonishing conclusion that we are "just a bundle of neurones". It is a classic example of the reductionist tendency to deny the reality of what has been given a scientific explanation. Note that this conclusion is drawn, even before the scientific explanation is actually available, simply on the basis of the dogmatic assumption that one will be produced.

There are also phenomena that indicate that consciousness may be less closely tied to the physical brain than we currently assume. The most important research here is that of Rupert Sheldrake who has demonstrated, for example, that dogs know when their owners are returning home, even at some distance, and when the timing of the return journey is unpredictable. He has also shown that people are aware that they are being stared at, even when the person staring at them is not in their visual field. Though I am persuaded by his evidence for these phenomena, I do not want to suggest that they are scientifically inexplicable. Rather, they seem to indicate that an adequate scientific theory of consciousness will be more subtle and complex than reductionists currently assume, and will involve the postulation of something like what Sheldrake calls an "extended" mind, i.e. a mind that extends beyond the boundary of the physical body.

Artificial Intelligence

The third arena we will look at in which reductionist ideology operates is artificial intelligence. Indeed, there is perhaps no other current area of science that is as heavily ideological as artificial intelligence. Though there is a wing of artificial intelligence work that has its eye on practical applications, for example helping doctors to combine large and complex sets of clinical data to arrive at diagnoses, much artificial intelligence has an openly ideological thrust.

The basic proposition is that the human mind is essentially a computer program. That is the assumption that makes artificial intelligence such an ideologically loaded and controversial area. In support of that basic proposition, efforts are then made to simulate on computers all intelligent human functioning. The hope is held out that it will be possible to develop a computer that can perform all intelligent tasks of which humans are capable. Indeed, the hope is that computers will be an advance on human beings, because they will be more dependable and not subject to illness and mortality.

Various arguments have been advanced for and against the basic proposition that the human mind is essentially a computer, though none has commanded general assent. Probably the most discussed argument has been the "Chinese room" argument of John Searle. He envisages a room into which messages are passed in Chinese. There is a very complex rulebook, which enables an appropriate answer in Chinese to be devised and passed out. However, that is all done without any understanding of what the messages actually say. Searle's point is that computers are in a similar position. Even when they are successfully programmed to give appropriate responses, they have no understanding of what the messages mean, in the sense that they have no knowledge of what in the world the messages refer to. Though a variety of refutations of Searle's argument have been proposed, none has commanded widespread acceptance, and it seems to me that the Chinese room parable makes a good point.

There is also an issue about the hardware on which cognitive programmes are instantiated. The silicon of which computers are made is radically different from the organic stuff of the brain, and Searle suggests that human intelligence can only be instantiated on the latter. That may well be so, though it is not clear that there is a convincing argument as to why it should be so.

It can be conceded that all intelligent, rule-governed cognitive activity can, in principle, be programmed on a computer. However, much human intellectual activity may not in fact be rule-governed, or at least the implicit rules that govern it may not yet be understood. There is also the gulf between simulation and replication. Even if their computers can simulate human intelligent performance, as in chess playing, they often do it in different ways from human beings, and in particular not with the same understanding of what they are doing. They simulate, but do not replicate.

Leaving aside such arguments, there is the practical matter of how much progress is being made in simulating human intelligence on computers. It should be accepted that such progress has been considerable, though it falls far short of the bold predictions that have repeatedly been made. It is conspicuous that computers lack anything corresponding to actual human experience. If the arguments of the previous section are correct, to the effect that qualia are scientifically demonstrable and have significant consequences for human performance, the lack of qualia in computers is a serious deficiency.

A telling illustration is how far it is possible to programme emotions in computers. In as far as the situations in which emotions arise are predictable, it is possible to programme a computer to have emotions that are appropriate in particular kinds of situations. It is also possible to programme some kind of emotional expression, at least for a computer to compose sentences that express emotional attitudes and to utter them through a voice box. In those limited ways, a computer can be programmed, for example, to simulate anger. However, a computer would have no understanding of what it was angry about, and no actual experience of anger.

Of course, these may this may not be insurmountable roadblocks. The development of parallel distributed processing in computer programming was a radical paradigm shift, and rendered things possible in programming that were difficult before, such as the simulation of learning by computers. There may be further paradigm shifts in computer programming ahead. So, it would be foolish to be dogmatic about what is and is not possible. However, for now, the gulf between human intelligence and the computer simulation of it looks to be unbridgeably wide.

If so, the claim that all intelligent human functioning can be simulated on a computer collapses, and so does the claim that the human mind is essentially a computer program. There is no doubt that the analogy between human cognition and computer operations has been a fruitful one, and provided a precise language in which to couch theories in cognitive psychology. However, progress in artificial intelligence gives no support to the reductionist claim that the human mind is nothing but a computer program.

● ● ● Conclusion

One puzzle about ideological reductionism is why it has such appeal. There is quite a sharp division between those who are excited by the possibility of explaining everything in reductionist terms, and those who find the idea naive and unattractive. Such different attitudes are bound to persist, and it is probably healthy for scientific debate that they should do so.

However, a clear distinction needs to be drawn between actual scientific research and ideological reductionism. Reductionist ideology is not a necessary precondition for a successful scientific research programme, neither is there anything in current research which justifies it as a conclusion. As an ideological position it is entirely detachable from scientific enquiry. My claim in this brief paper is that science would be better off without it. Abandoning reductionist ideology would be better both for open-minded scientific enquiry and for the public perception of science.

THE LIMITS OF BIOLOGICAL REDUCTIONIST EXPLANATIONS OF THE HUMAN CONDITION

Steven PR Rose

Steven Rose is a neurobiologist and professor. He chairs the Open University's Department of Biology (UK). He has published 9 books and more than 200 neuroscience research papers. Through the Brain and Behaviour Research Group he established, Professor Rose has focused his research on understanding the cellular and molecular mechanisms of learning and memory. His efforts in this area have led to the publication of some 300 research papers and various honours from several countries including Russia and the Netherlands.

eductionism is a portmanteau word; its multiple meanings generate confusion and raise hackles. Reductionism's advocates regard it as the one certain path to scientific knowledge; its opponents may concur, but see this knowledge as providing at best an impoverished and at worst a misleading understanding of the world. The issues are methodological, philosophical and ideological.

Reductionism as Methodology

The living world is characterised by dynamic complexity. But we find it easier to understand phenomena, whether the behaviour of molecules or of animals, if we can isolate them from the rest of the world and alter potential variables singly. It is hard to make sense of what you observe if several features of a system are changing simultaneously. Reductionist methodology simplifies, and enables one to generate seemingly linear chains of cause and effect. It is no surprise that it has been both powerful and attractive over the last 300 years, providing unrivalled insights into the physical and chemical mechanics of the universe, because it often seems to work. Experiments are productive, findings replicable, predictions about the world are confirmed.

But living systems are not simple. They involve many interacting variables. Parameters are not fixed. Properties are non-linear. And the living world is highly non-uniform; the exception is nearly always the rule. So if one is not careful, the simplifying constraints which the methodology offers soon cease to be helpful supports to theory, and instead become straightjackets. What happens in the test-tube may be the same, the opposite of, or bear no relationship at all to what happens in the living cell, still less the living organism in its environment. It all depends.

Science needs to simplify, to design experiments in which nature is caged, parameters held constant and variables changed one at a time. In studying an enzyme reaction for instance, one might hold the acidity of the solution constant and change the temperature, or vice versa, and derive simple equations describing the consequences. But if both change simultaneously, as indeed may happen in "real life", non-linearity results; the equations will not work and we will have lost our capacity to predict. Many of the phenomena science wishes to explain, from the dynamics of weather and ecosystems to the history of life itself, from the orchestrated metabolism of a single cell to development of the fertilised egg into the fully formed adult and the workings of the human brain, seem irreducibly complex. Isolating single variables can only confuse, and new scientific methods capable of dealing with complexity are required.

●●● Philosophical Reductionism

Classically the goal of science is to find all-embracing explanations –

preferably couched mathematically – of seemingly disparate phenomena. Reducing the fall of an apple and the rotation of earth and moon to examples of gravitational forces is an example. So – the classic case – is the recognition that the evening star and the morning star are both manifestations of a single planet, Venus. But this does not work too well in biology. Despite claims to the contrary, "a gene" is not straightforwardly identical to "an array of nucleotides within a strand of DNA", and the more we learn of molecular biological mechanisms, the less such a simple equation can be seen to hold.

The philosophical reductionist view is that because science is unitary, and because physics is seen as the most fundamental of the sciences, then an ultimate Theory of Everything will be able to reduce chemistry to a special case of physics, biochemistry to chemistry, physiology to biochemistry, psychology to physiology, and ultimately sociology to psychology and hence to physics. As James Watson put it with characteristic brutal clarity: "There is only one science – physics; everything else is social work." But is such a reduction really even theoretically possible? Consider the relationship between physiology and biochemistry, exemplified by the study of muscle. Physiology studies muscle contraction, biochemistry the molecular processes that occur during this contraction. The biochemistry of this process is pretty well understood down to some of the minutest molecular details. So why cannot we just replace the physiologist's statement about muscle contraction with a statement about the proteins that constitute the muscle?

If the purpose of doing so is to claim that the biochemistry is causally responsible for the physiological event, this is a very different use of the word *cause* from the way it is normally employed to describe a relationship in time between cause and effect, in which the proximal cause of the muscle twitch is provided by the physiological description of impulses travelling down a motor nerve to the muscle. The biochemical process does not *precede* the muscle contraction; it "describes" the muscle contraction. We are really making not a causal but an "identity" statement. But the physiological and biochemical descriptions serve quite different and complementary purposes. The biochemistry of the muscle twitch does not occur in the isolation of a test-tube. Muscles are anatomical structures, with specific locations within the organism. The relationships between molecular processes are organised in space and time in a manner that is not implicit in their chemistry. The *meaning* and *function* to the organism of the muscle twitch is apparent in the physiology and anatomy, but quite absent from the biochemistry. It cannot simply be eliminated.

Each "level" of complexity of nature involves new interactions and relationships between the component parts, that cannot be inferred simply by taking the system to pieces. Yet philosophical reductionism implies that even if higher order properties are emergent they remain secondary to lower order ones. The lower the order the greater the primacy. Parts come before wholes. The nature of evolutionary and developmental processes in biology means that there is no such necessary primacy. Wholes, emerging, may in themselves constrain or demand the appearance of parts.

The enthusiasm for reductionist explanations says more about the training and mind-set of scientists than it does about the nature of the processes and phenomena we study. We may – I believe we do – live in a world which is ontologically unitary, but to understand it we need epistemological diversity. Certain phenomena are only manifest, and only interpretable at a particular level of organisation of the world. I study memory, and can recount in great detail the molecular and cellular cascade of processes occurring in the brains of my experimental animals when they learn and remember a particular task. But however fine-grained and complete my biochemical analysis is I do not expect to find brain memory embedded in the molecules. It is a property of the system of interconnected nerve cells, and the organism in which those cells are located, to which the biochemical processes pertain.

Reductionism claims simplicity, and whilst I do not doubt that neuroscience will in due course be able to specify with a fair degree of accuracy exactly those cells and brain processes which become engaged when a person is angry or in love, I cannot conceive of a time in which we would wish to replace the meaningful statements "I am angry" or "I am in love" with a statement about the position and firing state of the hundred billion neurons in my brain and the hormones and neuromodulators circulating between them. The relationship between neurobiological statements about the brain and those about mental states is not one of causation in the sense that the first is primary and the second merely an epiphenomenon, but of translation – two different and equally legitimate languages in which to describe the same phenomenon – as when the same furry brown purring object is called cat in English and *gatto* in Italian. Which language we decide to use should be decided by context, not philosophy. Despite Watson, biologists need to grow out of their physics-envy.

● ● ● Reductionism as Ideology

Which brings me to the final issue: *reductionism* as ideology. By this I mean the seemingly irresistible tendency over recent years to seek to explain and indeed "treat" complex social phenomena in terms of disordered brains or genes. The list of such putative explanations is long: genes are supposed to be responsible for violence, "'antisocial and criminal behaviour", "impulsivity", non-heterosexual orientation, alcoholism, "female intuition", male rape, adultery, infanticide, and even "compulsive shopping". This ascription of unilinear causal power to genes, superseding all other levels and forms of explanation, has several peculiar features. First, the selective choice of behaviours: I am not aware of research on genes for bank fraud or homophobia. Second, the evangelical zeal with which such claims are made by the researchers concerned and greedily reported in the press, which takes an almost prurient delight in talking dirty. Third, the neat way in which such genetic claims to locate the causes of social problems within individual biology fits the needs of societies, like those of both Britain and the US, which have largely abandoned the belief in the possibility of achieving a more just and egalitarian society through political action.

It no longer becomes appropriate to seek the causes of violence on the streets of the US in terms of poverty, racism or the ubiquity of handguns. Instead, research programmes are directed towards locating abnormal biochemistry, itself presumed to be genetically caused, in the brains of inner city infants, which may "predict" an individual destined to violence and anti-social behaviour. Similarly, the presence of vodka-sodden drunks on the streets of Moscow prompts a major Russian research effort into the molecular biology of alcoholism. If Karadic and Mladic ever arrive in The Hague they may even plead abnormal monoamine oxidase activity in their brains as the cause of Bosnian genocide.

All of which might be dismissible if the genetic claims could be shown to be "good science". However, even by their own standards and despite publication in journals of the highest repute, they rarely are. Tentative findings are trumpeted, retractions and failures to confirm (as time after time in the case of "schizophrenia genes") are received in silence. The much publicised claims to have discovered a gene "for aggression" are based on the study of eight men over three generations within an extended Dutch pedigree, whose crimes varied from violent temper to exhibitionism, rape and arson, all of whom carried a genetic "marker" for a particular enzyme of neurotransmitter metabolism. Would that I could get away with publishing a paper based on the study of only eight animals! There is a reductionist cascade in such an approach at each point of which fallacious thinking can and often does appear.

Reification converts a dynamic process into a static phenomenon, a phenotype. Violence is the term used to describe dynamic processes of interactions between persons, or even between a person and their non-human environment. Reification transforms the process into a fixed thing – aggression – which can be abstracted from the interactive system in which it appears and studied in isolation. Yet if the activity can only be expressed in an interaction between individuals, to reify the process as if it were an isolable character attached to an individual is to lose its meaning.

Arbitrary agglomeration carries reification a step further, lumping together many different reified interactions as if they were all exemplars of the one character. Thus *aggression* becomes the term used to describe processes as disparate as a man abusing his lover or child, fights between football fans, strikers resisting police, racist attacks on ethnic minorities, civil and national wars. Agglomeration proceeds by assuming each of these social processes is merely a reified manifestation of some unitary underlying property of the individuals, so that identical biological mechanisms are involved in, or even cause, each.

As with each step in the reductionist cascade I am describing, the problem lies not in the fact that as researchers, within the methodology available to us, we need to classify or group different types of observation as belonging in some way together. These are not inevitably illegitimate steps. Science often proceeds by alternately grouping together different phenomena as aspects of the same (lumping) and recognising differences between them (splitting). Lumping is however inappropriate as applied to "violence", as in these examples. Grouping arson and exhibitionism in the same category, as in the Dutch pedigree mentioned above, is not likely to make much sense to either a criminologist or a judge and jury in court. To get round this difficulty, some researchers have recently relabelled these cases, so that they no longer appear as examples of "violence", but of a different category, of "anti-social behaviour". But such relabelling only makes it worse. Just as agglomeration lumps disparate activities, so the identical act may be regarded as socially acceptable or unacceptable depending on the circumstances. Bombing a government building if you are a pilot and your nation is at war with those you are bombing is socially praiseworthy; on the other hand if you are a national of the society whose buildings you bomb you are guilty of the anti-social behaviour called terrorism.

Improper quantification argues that reified and agglomerated characters can be given numerical values. If a person is violent, or intelligent, one can ask how violent, how intelligent, by comparison with other people. This assumption that any phenomenon can be measured and scored is reflective of the belief that to mathematicise something is in some way to capture and control it. The best-known example is IQ. "Intelligent behaviour", essentially an interactive process between an individual and others, or with the social, living and inanimate world, becomes fixed as a unitary character (*reification*). Many different examples of such behaviour are then all taken to be manifestations of something called, as if finally to freeze dynamics into statics, "crystallised intelligence", and given a special symbol, g, originally introduced by Spearman in the 1920s. Tests are then devised to measure this inferred hidden constant. The extraordinary belief is that all the multiple aspects of behaviour that go to comprise what we may recognise as intelligence – speed and accuracy of responding to new information, skill at deriving meanings from ambiguous social situations, capacity to innovate in novel environments, and many others as well – can all be reduced to a single number, so that the entire human population can be ranked, just as they might be if we were to line them all up by height.

Belief in *statistical normality* assumes that in any given population the distribution of such behavioural scores is Gaussian. Again the best-known example is IQ, the tests for which successive generations of psychometricians refined and remoulded until it was made to fit (almost) the approved statistical shape. That is, tests which did not result in distributing the population according to the curve were rejected or modified, until they fitted the curve, a feat achieved between the wars in the various revisions of what became known as the Stanford-Binet, originally developed in the 1920s. Yet the assumption that the entire population can be normally distributed along a single dimension is to confuse a statistical manipulation for a biological phenomenon.

The power of this reified statistic should not be underestimated. It conveniently conflates two different concepts of "normality". The statistical sense of the term does not have a "value" attached to it, it merely describes a particular shape of curve which has the property that 95% of its area is to be found within two standard deviations of the mean. But in common parlance the term normal is indeed normative. It describes not merely how things are, but how they ought to be; to lie more than two standard deviations from the mean in a Gaussian distribution is to be "abnormal".

Spurious localisation means that having reduced dynamic processes to reified numbers, they cease to be a property even of the individual, but instead becomes that of a part of the person. Thus arises the penchant for speaking of, for example, schizophrenic brains, genes or even urine, rather than of brains, genes or urine derived from a person diagnosed as suffering from schizophrenia. Of course, everyone knows that this is a shorthand, but the resonance of "gay brains" or "selfish genes" does more than merely sell books for their scientific authors; it both reflects and endorses the modes of thought and explanation that constitute neurogenetic determinism, for it disarticulates the complex properties of individuals into isolated and localised lumps of biology.

Thus recent years have seen an unusually polemical debate, more reminiscent of the early days of 19th century phrenology than of modern research, amongst different neuroanatomists each claiming to

101

have found the brain seat of homosexuality. Two regions in particular have contended for the honour of conveying male same-sex preference, the corpus callosum, and the hypothalamus. The structure of the argument deployed by those seeking to locate homosexuality in a bit of the brain or an aberrant gene shows all the features I have already described for violence and intelligence, and more besides. The expression of same-sex preference is scarcely a stable category either within an individual's lifetime or historically – indeed, that it might be used as a term to describe an individual, rather than part of a continuum of sexual activities and preferences available to all, seems to have been a relatively modern development. What the reductionist argument does is to remove the description of sexual activity or preference as part of a relationship between two individuals, reifies it and turns it into the phenotypic "character" resulting from one or more abnormal brain structures. As always, it deprives the term of personal, social or historical meaning, as if to engage in same-sex erotic activity or even to express a same-sex preferred orientation meant the same in Plato's Greece, Victorian England or San Francisco in the 1960s.

Misplaced causation refers to the point where neurogenetic determinism introduces its misplaced sense of causality. During sexual or aggressive encounters people show dramatic changes in, for instance the levels of circulating steroid hormones and adrenalin in their bloodstream and the release of neurotransmitters in their brains, all of which can be affected by drug treatments. People whose life history includes many such encounters are likely to show lasting differences in a variety of brain and body markers. But to describe such changes as if they were the causes of particular behaviours is to mistake correlation or even consequence for cause. When you have a cold, your nose runs. Yet despite the invariable correlation of the two, it would be a mistake to believe that the cold was caused by the nasal mucus; the chain of cause-effect runs in the reverse direction. Nor, despite the fact that Prozac both inhibits serotonin re-uptake mechanisms and may increase the likelihood of you committing suicide or murder, does this mean that the level of serotonin release in your brain is the cause of your desire to kill yourself or someone else. After all, when one has toothache one can alleviate the pain by taking aspirin, but it does not follow that the cause of the toothache is too little aspirin in the brain. This problem has dogged interpretation of the biochemical and brain correlates of psychiatric disorders for decades, yet it still continues.

Dichotomous Partitioning. If aggression, anti-social behaviour, or homosexuality are "caused" by some "abnormality" in brain structure or biochemistry or hormonal imbalance, what "causes" these in their turn? They could of course be the consequences of some feature in the environment. More often, though, attention turns to those well-known first causes, the genes, and the apparatus of heritability studies is wheeled out. For even if there is difficulty in regarding such socially defined attributes as simple phenotypes, if they correlate with a "real" measure such as the level of an enzyme or neurotransmitter, then the heritability of this can surely be determined. A good example of this mode of thinking is the claim that IQ test scores correlate with a more neurophysiological measure referred to as "inspection time", whose heritability can then be assessed. The fact that a heritability measure is rarely applicable to the human situation, is widely misunderstood and in most cases meaningless, has not prevented behaviour geneticists and psychometricians endeavouring to apply it, nor deprived it of its ideological resonance, as when it is reiterated that the heritability of intelligence – or rather of IQ test score – is as high as 80%. Political orientation, neuroticism and attitudes to military drill, royalty, censorship and divorce, amongst many others, are all supposed to show relatively high heritability. Indeed it becomes hard to find any human attribute or belief, even the most seemingly trivial, to which the heritability statistics fail to yield significance. Newly sophisticated statistical techniques are employed purporting to show that even those conditions for which major genetic causation cannot be shown are in fact the result of the small additive effects of many genes. (The sad truth is that there is no distribution of a phenotype to which, given enough assumptions about partial penetrance and incomplete dominance, a genetic model cannot be fitted). And whilst no-one claims that heritability equals destiny, nor that the figure provides information about any specific individual rather than measures variance within a population, nonetheless, the whole tenor of the approach is to transfer the burden of explanation, and if appropriate of intervention, from the social or even personal level to that of pharmacological or genetic control.

Confounding metaphor with homology. If first causes are genetic, it then becomes appropriate to seek for equivalents of the human behaviour under consideration in the non-human animal world; that is, to find an animal model in which the behaviour can be more readily controlled, manipulated and quantified. Place an unfamiliar mouse into a cage occupied by a rat and the rat is likely eventually to kill the mouse. The time taken for the rat to perform this act is taken as a surrogate for the rat's aggression; some rats will kill quickly, others slowly or even not at all. The rat that kills in 30 seconds is on this scale twice as aggressive as the rat that takes a minute. Such a measure, dignified as muricidal behaviour, serves as a quantitative index for the study of aggression, ignoring the many other aspects of the rat-mouse interaction, for instance the dimensions, shape and degree of familiarity with the cage environment of the participants in the muricidal interaction, whether there are opportunities for retreat or escape, and the prior history of interactions between the pair. These are not merely speculative variables; many have been studied in detail by ethologists and shown to affect profoundly the nature of the relationships between the animals. But the reductive procedure goes further, for it then assumed that, just as time to kill becomes a surrogate for a measure of aggression, so this behaviour in the rat is transmogrified into an analogue of the aggression shown by drive-by gangs shooting up a district in Los Angeles. That is, if one can find physiological or biochemical mechanisms - brain regions, neurotransmitters or genes – associated with the so-called aggression in mouse-killing rats, then there should be equivalent or identical brain regions, neurotransmitters or genes involved in human "aggression". This type of evolutionary fantasy at best confounds a metaphor or analogue with a homologue. At worst it simply makes a bad pun on different meanings of the word aggression. But it has become the vital, ultimate link in the chainmail armour of reductive ideology.

The Consequences of reductionist Fallacies

Methodological reductionism has proved a powerful and effective lever with which to move the world. We owe to it many of the most penetrating insights into mechanisms in every field of science, including biology. But especially in biology, complexity and dynamics, open rather than closed systems, are norms rather than exceptions, and the methodology of reductionism however powerful, has difficulties in dealing with complexity. Indeed it may be positively misleading.

Reductionism as an ideology, insisting on trying to account for higher-level phenomena in terms of lower-level properties, hinders biologists from thinking adequately about the phenomena we wish to understand. But two consequences at least lie in the social and political domain rather than the scientific. Reductionist ideology serves to relocate social problems to the individual, thus "blaming the victim" rather than exploring the societal roots and determinants of the phenomena that concern us. Violence in modern society is no longer to be explained in terms of inner city squalor, unemployment, extremes of wealth and poverty and the loss of the hope that by collective effort we might create a better society. Rather, it is a problem resulting from the presence of individual violent persons, themselves violent as a result of disorders in their biochemical or genetic constitution.

The second immediate social consequence of reductionist ideology is that attention and funding is diverted from the social to the molecular. If rates of alcoholism are catastrophically high in the former Soviet Union or amongst first Americans or Australian aborigines, the ideology demands funding research into the genetics and biochemistry of alcoholism. And it becomes more productive to study the roots of violent "temperament" in babies and young children than to legislate to remove handguns from society. The point is that, as the whole of my argument up till now has stressed, for any phenomenon in the living world in general and the human social world in particular, one can offer multiple forms of explanation. But for any such phenomenon there are also "determining levels" of explanation: those which most clearly account for its specificity and point to potential sites of intervention.

So whilst in an ontologically unitary universe it is axiomatic that there is something different about the biochemical and physiological state of someone who is in the process of committing a murder from those states in the same person when he is in a prison cell, and probably between the murdering individual and someone who in similar circumstances does not murder, this difference cannot be relevant to answering questions about the causes and responses to social violence. Nor, therefore, can it represent the appropriate level at which to intervene if we wish to reduce the amount of violence on the streets. A programme devoted to the detection of what levels of serotonin might predispose a person to an increased statistical possibility of engaging in one of a number of activities, from suicide through depression to murder, followed by the mass screening of individual children to identify at-risk individuals, their drugging throughout life, and/or raising in environments designed to alter their serotonin levels, which is, after all, the action programme that would result from an attempt to define the genetic/biochemical as the right level for intervention, only has to be enunciated to demonstrate its fatuity. Good, effective science requires a better recognition of determining explanation and hence the determining level at which to intervene. Failing this it becomes a waste of human ingenuity and resource, a powerful ideological strategy of victim-blaming and a distraction from the real tasks that both science and society require.

The dangers of such reductionism are manifest: a Gresham's law in which bad science drives out good, a misapplication of scarce scientific resources, the search for technical fixes to complex social problems, the belief that the world of living systems is simple and not complex and that, ultimately, as living humans we are not in charge of our destiny but merely lumbering robots serving the needs of our genetic puppet-masters.

CREATING CONTEXTS:
REDUCTION OR EXPANSION?

Marilyn Strathern

Marilyn Strathern has been a professor of social anthropology at Cambridge University (UK) for nearly 20 years. Her field work in Papua New Guinea and England has delved into diverse issues of gender, kinship and new reproductive technologies. Professor Strathern has published numerous articles and books, including 'Women in between' and 'Technologies of procreation'. Professor Strathern is an honorary foreign member of the American Academy of Arts and Sciences. Her latest area of interest and research regards intellectual property issues.

It is ironic that in one of the places where the public in Europe or North America readily assents to the interventions of science, they should get it so wrong. Genetic essentialism, the notion that genes may be code for particular elements of human behaviour, can after all be interpreted as society giving the biological sciences their due. Whatever other attitude people have, they are ready to believe that science can unfold facts about the world. And whereas facts about the environment or about the beginning of the universe may be treated with some scepticism, it would seem that facts about *human nature* get absorbed very easily. Indeed, there is apparently a great appetite for them. If explanations reduce complex interactions to simplistic notions about genes, the explanations themselves seems to mushroom everywhere – and often with great ingenuity.

There are some lessons to be learned from this. This is a very brief comment on some recent thinking about the nature of public reactions; it focuses on the way knowledge proliferates. The subjects sketched in here are "people" not otherwise identified except as belonging to a broad cultural field one could call "Euro-American", and "researchers" who come from the same population. Any specific perspective of mine is likely to be British.

● ● ● Objects of Knowledge

A social fact about the life sciences, then, is people's readiness – or need – to take on board information about their genetic make-up.
This is partly but not entirely due to the rationalist presumption that one should act on the knowledge one has. (People generally regard it appropriate that government policies should seek "evidence-based" approaches to medical provision.) Ideas about human nature also feed into the knowledge on which people base their own daily routines of behaviour, their judgement of others, attitudes to socialisation, care of themselves, and aspirations for their children. The underlying connection is the extent to which knowledge of biology is built into the way people think about the make-up of themselves as human beings. One small example of considerable interest to social anthropologists concerns families and kinship.
Kinship, in this view, is based on the natural facts of procreation. Although there are many types of relationships that appeal to social arrangements (the duty of care of adopting parents, the obligations that arise between those who live in the same household, gay parentage), they are often modelled on biological ones. Moreover, while not all social arrangements are seen to stem from biological connections, biological connections cannot be discovered without having implications for those arrangements. Genetic connection is a prime example. So, as was said a long time ago[1], anything that science discovers about the biology of procreation will be taken as evidence of kinship.

1 David Schneider.

If in these procedures kinship is validated, then so too is biological science. For Euro-Americans, there is always more to learn about themselves and about how they are related. And that is because, as Jeanette Edwards has consistently argued, people already have plenty of ideas about the nature of their connections to one another. Genetic knowledge falls into a field of assumptions about the significance of the transmission of traits, medical fortunes, kinship identity, parenting, family relations, and so forth. This is a form of reductionism not dissimilar to scholars' appeal to the discipline that has already furnished them with frameworks for knowledge.

In other words, there is an important reductionist tendency built into people's readiness to understand. To the observation that "human behaviour is always embedded in a pre-existing social environment, and most of it is channelled through cognitive processes and cultural idioms", we can add that the same is true of people's modes of explanation. Whether they want to press their knowledge into use, or else have no option but to take into account "the biological facts", new knowledge gets channelled into old suppositions, above all about procreation and how people come to "share genes".

For here of course they bring wisdom of their own. Edwards, in her writing on the readiness of the residents of a town in Lancashire in the UK to talk about the new genetics and assisted conception procedures, comments on the way that they take knowledge to make knowledge. If, as people do everywhere, they translate information into its impact on the experience of their own and other people's lives, it is important that the observer does not reduce that to passive consumption. Indeed Edwards contributes to debates that have, over the last half decade, accompanied the UK policy shift from promoting the "understanding" of science (a deficit model – all one needs to address is people's ignorance) to promoting "engagement" with it (a democracy model – people as active participants in the world around them).

111

For here of course they bring wisdom of their own. Edwards, in her writing on the readiness of the residents of a town in Lancashire in the UK to talk about the new genetics and assisted conception procedures, comments on the way that they take knowledge to make knowledge. If as people do everywhere they translate information into its impact on the experience of their own and other people's lives, it is important that the observer does not reduce that to passive consumption. Indeed Edwards contributes to debates that have over the last half decade accompanied the UK policy shift from promoting the "understanding" of science (a deficit model – all one needs to address is people's ignorance) to promoting "engagement" with it (a democracy model – people as active participants in the world[2].

> Nina: "I did once read that some children were slower through eggs having been frozen. Children from frozen embryos were slightly slower. Now I only read it once. They [journalist, scientists] seemed to just put it out and then let it go. And it said they [the children] caught up, but it was harder for them to catch up to other children. So it [embryo technology] must have *some* impact." (original emphasis)
>
> Rose: "I didn't agree with that one [media report] about that woman who wanted [the] baby with her husband after he had died … I don't think it's fair on the child … it would know that its dad was dead and it would never get to see him."
>
> Nina only had to read the news item once, note, for the information to make an impact on her.

The people Edwards engaged in talking about the new reproductive and genetic technologies brought to their joint discussions much information about what they expected other people's reactions would be. Rose could mind-read the prospective child. Others put thoughts

---

2 NOWOTNY, H. et al., *Re-Thinking Science Knowledge and the Public in an Age of Uncertainty*, Cambridge, Polity Press, 2001 ; Andrew Barry (the scientific citizen).

into the minds of prospective parents: how they would behave, for instance, in reaction to the possibilities of "genetic testing". This led her to observe that the notion of scientific literacy is too narrow a concept to capture both the knowledge that people draw on to make sense of innovation in science and technology and "the objects of knowledge they create in so doing[3]". (my emphasis)

And much of that information was already – it had to be – in their repertoires. Though invariably refashioned through comparison with experience, exploration of concepts, and so forth, it is their own objects of knowledge that are re-invigorated, worked on and transformed.

But what is this reductionism? In another sense, these interpretations are the very opposite – they are also non-reductive. One could look at it quite differently and say that people are putting scientific, including biological, knowledge into *diverse contexts*. They are actually expanding the sphere of application, finding new relevance. The rush to accept the idea that there is a genetic basis to many kinds of dispositions and patterns of behaviour suggests the topic opens up innumerable contexts. People are willing to jump from one domain of social life to another, to make connections, draw analogies, to leap from single-gene medical conditions to complex behaviours, while all the time imagining they are still in the same logical frame. The problem for the biologist is precisely that such proliferations run away with the facts – they are not kept as observations that belong only to internal scientific debate, but people want to draw from them lessons about themselves. Perhaps even genetic essentialism also reflects an appetite to know more.

Edwards' acquaintances in Lancashire were not necessarily genetic

3 EDWARDS, J., "Public understanding of science: deficit and democracy", paper delivered at session on *Science and Democracy: the Open Society Revisited*, convened by Nigel Rapport, British Association for Advancement of Science, Festival of Science, Glasgow, 2001. For a fuller ethnographic background, see EDWARDS, J., *Born and Bred: Idioms of Kinship and New Reproductive Technologies in England*, Oxford, Oxford University Press, 2000.

essentialists. But in drawing debates about genetic testing (say) into their own experiences, they multiplied the number of contexts in which the information could lodge. And they were doing much more than talking just about families and parenting. They were concerned about regulation, for example, and drew the analogy between scientists and "the government": you have to keep questioning or else the elite just steamroll ahead and produce the kind of society they would like for themselves.

She concludes[4]:

> "Not to broaden the concept of 'scientific literacy' in ways which can include how people identify wider social, political and economic aspects of scientific practice would be partisan, but so too would be using the concept as a catch-all to cover the diversity of ways in which people know the world."

In other words, the discussions had as their aim not just engaging with science (in the manner that present-day policy would encourage) but fuelling other parts of their lives, where issues arising from scientific practice join other components of background knowledge[5].

One of the characteristics of social knowledge that we need to grasp is its simultaneous reductive and expansive tendencies. Let us look more closely, then, at the whole question of context.

● ● ● Context

Knowledge is not an organism. The purport of this will be clear in a moment.

---

4 EDWARDS, J., 2000, *op.cit.*
5 The jargon I use for this process is "merographic relations". In Euro-American knowledge systems, every whole is also a part of another whole; anything can be put into some context or other, but nothing is completely encompassed by any one of them and may be found in other contexts too. Otherwise put, Euro-Americans make sense of things by describing them as part of something else. STRATHERN, M., *After Nature: English Kinship in the Late Twentieth Century*, Cambridge, Cambridge University Press, 1992.

As we have seen, the same instrument, contextualisation, may be both restrictive and expansive. The reductionism is clear. Summoning any one specific context restricts by its degree of specificity (though may acquire greater explanatory power thereby). The context reduces the scope of the enquiry by whatever is taken to be appropriate to it. Indeed, it may be regarded as scholarly and rigorous to define a context in as restricted terms as possible. A significant factor here is that context reduces information by the simple fact that foregrounding any one framework is to put others into background. In other words, when it comes to explanations for things, contexts compete for attention. The rest of this section turns to the proliferating potential of any one context to generate others.

The residents of the Lancashire town with whom Edward's spoke may have mobilised several parts of their lives to reflect on scientific innovation. Social scientists are often explicit about doing this. As a matter of method, one of the ways in which they attempt to "avoid" reductionism is to seek to put phenomena into "broader contexts". This is part of the knowledge-making process. At the same time knowledge itself must be treated as a phenomenon that needs to be put into context. A common way of doing this at the turn of the 20th-21st centuries is to think deliberately about the "implications" of this or that data or discovery.

For the social scientist, "context" is likely to signal "society" (or culture) at its broadest reach. The context is held to be rather like an environment. In a crude version of this, the environment is external or outside and thus envelops the phenomenon; a more sophisticated version sees the phenomenon electing to speak for the relevant environment as an organism creates its own ecological niche[6]. But in either respect the intention of "putting things into context" is to raise one's sights beyond the immediate phenomenon. In other words, the

115

6 Tim Ingold.

expectation is that increased information can be brought to bear on the issue or problem at hand, that is, bring about an increase in explanatory power.

A context is commonly used as an expansive instrument, then, precisely in order to counteract reductionist tendencies, and to bring in extra information. On the one hand, the context may seem "natural" to the phenomenon in question, but will enlarge or enhance everything one needs in order to understand it. An example is the recent public debate in the UK, and elsewhere, over patenting the genome. One response was the UK Nuffield Council's 2002 discussion paper on *The ethics of patenting DNA*. Not exclusively by any means, but very largely, this puts the debate into the context of legal presumptions and practices. Patents are a legal category in the first place, so we can say that this exercise in contextualisation was partly moulded by the problem under discussion. The law was a natural niche in this respect.

On the other hand, contextualisation may entail bringing in quite different orders of data. The orders may be determined by overt political or social interests, which then come to seem extraneous to the phenomenon itself. People may argue that a gene itself is not a social fact, but how information about it is relayed, how it is publicly known, and the implications that knowledge will have, all certainly are, and it is only right that the public's views are voiced. Examples abound in ELSI (Ethical, Legal and Social Implications) programmes. When it is thought right that the problem or issue under discussion should be scrutinised for its wider implications, information is sought from innumerable sources.

Now I have to use the word phenomenon in order to cover entities such as objects of knowledge or explanations. Thus the gene as an object of knowledge constitutes an observation, a set of facts, a recording, a representation even. Despite the parallels, we are not in fact dealing with an organism that has a life to lead. An organism's interaction with the environment is intimately bound up with the life-cycle of them both, and is thus specific to them both, to the extent that one may refer to *its*, i.e. the organism's, environment. An object of knowledge, on the other hand, could be put into any one of innumerable other contexts and happily reside there, so to speak. It would seem that our capacity for increasing knowledge is confined only by the number of contexts we can reasonably take into account. Of course, what is widened out can always be narrowed down again: thus our capacity for increasing usable knowledge will be confined by the policy requirements that call it forth in the first place.

These observations are stimulated by Marcus Schlecker and Eric Hirsch[7]. Examining strands of an intellectual history in both media and cultural studies, and science and technology studies, they explicitly comment on scholarly practices of contextualisation. In particular, they observe the fashion in these disciplines for ethnographic methods that were held to enlarge various objects of knowledge by broadening the contexts in which they were appraised: "ever more contexts could be could be combined to garner knowledge and thus increase insight[8]." At the same time, contextualisation of this kind also had a reductionist role to play, in controlling the scope of enquiries. It would come to terms with the specificities of real life, with concrete locations and records of actual encounters.

In media studies, for example, this was part of an attempt to get to grips with the kinds of audiences that the media were reaching. Researchers began deploying ethnographic methods in order to discover the different ways in which media messages could be read. They then began tracing their subjects' lives beyond the domestic setting in which they were consumers of media products. At the same

---

7 SCHLECKER, M. et HIRSCH, E., "Incomplete knowledge : ethnography and the crisis of context in studies of the media, science and technology", *History of the Human Sciences*, 14 (1), 2001, pp. 69-87.
8 SCHLECKER, M. et HIRSCH, E., 2001, *op. cit.*, p. 76.

time, science studies turned their detailed attention to scientific biographies, and the attempt to capture particular contexts such as the lab culture. In turn, scientists' work was then followed beyond the laboratory, and investigation widened to include a great variety of artefacts, such as research equipment, technologies, science policy, and so forth. The interplay between reductionism (specifying particular arenas in which observation could be made) and expansionism (adding more and more frameworks for knowledge) no longer seemed under the researchers' control. It was reductionism that was leading to proliferation! Schlecker and Hirsch write:

"The shift to an ethnographic perspective entailed a kind of fractal effect. Instead of limiting the number of relational contexts, the narrowing down of research scope brought out yet more contexts. It was recognised that instead of bringing the researcher closer to his or her subject matter, the ethnographic perspective was opening up further domains[9]."

They were chasing the boundless multiplication of contexts. Their vocabulary is deliberate, and they refer to a fractal unfolding of complexity.

Obviously scale does not reside in phenomena, it resides in the tools by which we describe phenomena. For the researchers described by Schlecker and Hirsch were never going to exhaust the possibilities of knowledge in narrowing down or opening up the contexts they constantly enrolled (though they might enhance what they already knew, refine it and make it more sophisticated). For while either move could be pursued as though it were knowledge-increasing (the narrowness of concrete records / the multiple horizons of different perspectives), the amount of detail that was left to be coped with often seemed to remain the same. When some areas of knowledge are eclipsed, detail from other areas rushes in to fill the gap; when detail

---

9 SCHLECKER, M. et HIRSCH, E., 2001, *op. cit.*, p. 76.

about phenomena get reduced to its generalisable essentials, it find itself in a universe of generalisations.

That the quantity of knowledge can hold itself constant regardless of what scale is involved must be a commonplace in the biological sciences: after all there is as much work for a molecular biologist to do as there is for an ecologist. The same is true of social configurations as it is of objects of knowledge. Whether one is dealing with a handful of people, or with global organisation[10], the interactions may be of an equal order of complexity.

In social life, it may not just be quantity that is held across scales, so too may values of various kinds, ideas and formulations. A notable example is the constantly regenerated interest among Euro-American societies in the relationship between nature and culture, discovery and invention, organic development and socialisation – the relationship is created over and again. Individual and society is a version of it. As a result, whether one is tracing a genetic history or watching thoroughbreds at the races or weighing up what to tell someone about their ancestry, there is a continuity of expectations in, a not-so dissimilar re-encounter with, the same value relationships. Wherever one is, one is in a roughly familiar situation. This is how culture works: realising one can operate across numerous situations of different orders of magnitude or social character. If social life did not work this way, then the recent book edited by Mark Mosko and Frederick Damon would lose its purchase[11]. Although it takes its materials from beyond Euro-American societies, it explores, among other things, the behaviour of what one might call cultural fractals, identifying key formula which, replicated, replicate cultural patterning. It is way of understanding the repetition of recognisable values across numerous contexts of social life.

---

10 Bruno Latour.

11 Mosko, M. & Damon, F. (Eds), *On the Order of Chaos: Social Anthropolog y and the Science of Chaos* (forthcoming).

But where have the life sciences gone in all this? I have wanted to put discussion about reductionism into a particular context, namely the context of other contexts of knowledge-making. If it seems that other interests have displaced that initial one, perhaps that is an effect of contextualisation itself.

●●● Conclusion

We started with the reductionism that leads to protests that biologists, among other scientists, create too narrow a view when they (for instance) focus only on the organism and not on the organism's "natural context". But this leads to a counter problem: how in turn not to render appeals to social phenomena reductionist for the biologist. A gesture towards "society", with its people, public, stakeholders and the like, appears an equally opaque sink for explanations. If there is one aim in this short paper, it is to make certain social phenomena, especially the way people explain things to themselves, as complex (possibly even as interesting) as biological knowledge is.

There is one radical caveat. Unlike the intimate and specific relationship between an organism and its environment, social contexts for human activity, including the work of biologists and including the effort that people put into trying to explain things, are not exclusive to one another or mutual in that sense. Social contexts belong to a world full of them, that is, other contexts. It follows that deciding the context "in which" to put certain sets of data requires something close to political will.

One task ahead is to work out how to be selective across contexts. That is, how to augment knowledge-making in order to gain insight or elucidation without either being reductionist (feeling that one has focused on only a tiny part of the whole) or expansionist (that one can, from any quarter, add information to information). The sheer complexity would be daunting but one might even venture into what we could call comparative contextualisation (or in current management jargon, "distributed" contextualisation). That is, instead of picking off individual topics and in each case assembling the arguments for that field, a focus which can of course work brilliantly in its own terms, one moves on to tackle that difficult fact of social life, viz. that contexts are not insulated from one another. Certainly it must be as much a commonplace in science research as it is in the humanities that developments move ahead across several fronts at the same time.

I think this also goes for crises and problems that demand interdisciplinary effort to resolve them[12]. Among the problems in scientific policy that the UK Government faced recently, for example, was the way in which arguments over GM crops and the authority of scientific data were affected by perceptions of the preceding BSE crisis. Now, it is "against" public "confusion" between such issues that researchers insist on separating out specific problems. Yet there may also be some mileage in comparing the crises. So instead of trying to pare the details down to their simplest in order to understand the essentials, to tackle things bit by bit, to undertake tasks of purification[13], one should admit the contamination. That kind of crossing of ideas across contexts, which is a confusion to the reductionist, works as one of the very facilities by which researchers enhance their own knowledge. It also works as a facility by which people will admit scientific knowledge into their own scheme of things.

---

12 CALLON, M. (Ed.), "An essay on framing and overflowing: economic externalities revisisted by sociology", *Laws of the Markets*, Blackwell Publishers.
13 Bruno Latour.

Life Sciences and Democracy

Science is by no means practised in a political vacuum. A number of failed political regimes during the last century (typically Soviet Communism and National Socialism) claimed to be based on scientific principles. Earlier in history, two major waves of European colonialism took place during the Renaissance and the Enlightenment, which coincided with key breakthroughs in science. Many see modern science, and life sciences in particular, as merely the most recent embodiment of the old ideologies that have been used to justify authoritarian actions.

Scientists have always insisted on claiming that freedom from value judgements is a central characteristic of science. Embedded within the Western science tradition is a doctrine of the free and unhampered pursuit of knowledge, based on a distinction between the understanding of the objective and "real" laws of nature on the one hand, and personal gain or social need on the other.

"Democracy" has never been a major issue in Western experimental science, as long as the resulting knowledge has not been applied to support a specific social doctrine. Yet in practice, findings from experimental biology have been used to reinforce the political ideologies of the time. Indeed, scientists themselves cannot escape the contemporary views of the world in which they live.

A number of new technologies that grew out of modern biology in the latter part of the 20th century created an explosive upheaval. An entire new industrial sector – biotechnology – arose; and subsequently a number of disciplines and medical practices started evolving into the "life sciences".

Today, many of the principles of socio-biology, social Darwinism in its various forms, eugenics and radical genetic *reductionism* – which suggests that a range of life choices may be already determined at birth – are readily embraced by ultra-liberals and people who believe unfettered globalisation is the best way forward for the world. As a consequence, both religious ultra-conservatives and anti-globalisation activists regard life sciences research with little sympathy, if not outright opposition.

How can modern science and scientists separate the good from the bad? How can we still pursue the noble quest for knowledge of the natural world (including ourselves) without being "captured" by political interests? What are the roles and responsibilities of life scientists in the social and ideological debates of our time? How meaningful is it to go on distinguishing between "basic" and "applied" science? How has the expansion of "biology-gone-global" affected its culture and basic values? And to what extent does the growing impact of biology on day-to-day life demand greater democratic guidance and public control of R&D in this field?

BEYOND UTOPIAS:
EVOLUTIONARY RATIONALISM AND NOOCRACY

Ladislav Kováč

Ladislav Kováč is professor of biochemistry at the Comenius University in Bratislava (SK). Professor Kovác's research focuses on bioenergetics – the study of the energetic aspects of life processes using biochemistry, physical chemistry and molecular biology. He is a founding member of *Slovakia's Academic Society* and the *Internationale Akademie Schloss Baruth*. He served as Czech education minister and as Czechoslovakia's Ambassador to UNESCO. Between 1990-1992, he was chairman of the UN Intergovernmental Committee for Science and Technology for Development. He has won many scientific awards, including the *Czech State Science Award* in 1985. He has written over 150 papers on biochemistry and genetics for international journals.

round Zero. That is the name that has been given to the empty spot in the centre of New York City, where the twin towers of the World Trade Centre collapsed under the terrorist attack on September 11, 2001. The present state of humankind has got in the spot its symbol and in its name a most concise characteristic[1]. Grand, grandiose is the knowledge of natural laws and on their basis the technological feats of humankind. Zero, or close to zero is the human knowledge of forces that direct human individual behaviour and social dynamics. Such forces motivated the actors behind the events in September 11, 2001. Unknown and uncontrolled, they continue to determine the behaviour of the main

---

1 Kováč, L., "Science and September 11th: a lesson of relevance", *World Futures*, 59, 2003, pp. 319-334.

actors in our times.

This is not to say that there has been a lack of effort to explain these forces. History abounds with countless speculations regarding the nature of both humans and society. *Ground Zero* may be conceived as an appeal to throw a fundamental doubt on these explanations: Have we been asking the right questions? Is it not high time to zero all our commonly accepted would-be explanations and to start anew? Has not the event of September 11, 2001 warned us that we have to revis the very foundations of contemporary thought? Is there not a most urgent task for natural sciences, primarily biology, to exploit all the attained knowledge in their field of enquiry so as to apply it to those problems that, until now have been the exclusive domain of cultural (human and social) sciences? The present study has resulted from such questions being posed.

● ● ● The Fallacy of European continental Thought: Belief in the unrestricted Potential of individual human Reason

European culture has one of its origins in Ancient Greece, 2,500 years ago. It started with abandoning an animistic interpretation of the world and attempting to explain it using naturalistic arguments – with the transition from Myth to Logos. With this transition, science was "born" on Earth. At the same time, this was the beginning of one of the main currents of European culture, of rationalism in Continental Europe, of a belief that the reason of human beings can, by means of logical inference, achieve an extensive and reliable knowledge of the world. Belief in unlimited power of individual reason may have reached its peak in the Enlightenment of Continental Europe in the 18th century (the English, Scottish and American Enlightenment had, with its empiricism, a different character), but it continues to persist not only in philosophy, but in science as well. Equating human consciousness with thinking and reasoning is still one of the dominant paradigms in cognitive sciences. Some prominent evolutionary biologists oppose the clumsy tinkering of biological evolution which, according to them, is "immensely stupid", with individual human's outstanding creative potential of conscious intelligence.

The situation is undergoing a substantial change. In cognitive sciences, the "affective revolution" is under way. No longer may thinking and reasoning, but the existential experience of perceptions and emotions, sensory and emotional qualia, be conceived of as the substance of consciousness. Human conscious deliberation appears to be just a "monomolecular layer" on the surface of an immense "ocean" of adaptive unconsciousness[2]. The latter represents knowledge which, progressively acquired by a ratchet-like mechanism of trials and errors, has been incorporated by biological evolution into the structure and chemistry of cells and organs and into unconscious evaluative and executive devices of the brain, and by cultural evolution into material artefacts and social institutions. "Reason", conscious reasoning, rationalisation seem to function mainly *a posteriori*, in order to explain and justify the actions that preceded the awareness of what was being done.

European faith in the power of individual reason has led unavoidably to contriving social utopias. Some utopia designers based their projects on the conviction that individual human reason has its parallel in the Reason of History, and hence on the belief in the existence of rational "historical laws" which should be revealed and exploited. Others maintained that cultural evolution, and within it the evolution of society, is blind, and even, like the biological evolution conceived by some evolutionary biologists, "immensely stupid", so that its products should be replaced and surpassed by much better

2 WILSON, T. D., *"Strangers to ourselves: Discovering the adaptive unconscious"*, Cambridge, Harvard University Press, 2002.

work designed by exceptionally intelligent individuals. Generating utopian projects has been a most cherished pastime among the brightest of European thinkers for centuries. The projects could not be materialised and, hence, neither approved nor falsified. Only the 20th century – "the century of science", as the saying sometimes goes – reached the capacity to provide technological means for trying out one such social utopias – Communism. Even though more than a decade has elapsed since the collapse of European Communism, this most important experience of humankind in the 20th century has not yet been properly evaluated. The principal conclusion from this lesson has not become commonplace: the failure of Communism dealt traditional European rationalism a decisive blow. In turn, it is promoting an alternative doctrine - evolutionary rationalism.

●●● The Failure of Communism: a cogent Case for evolutionary Rationalism

The theoretical basis of Communism, Marxism, was a logical, and probably unavoidable, culmination of rationalism in Continental Europe (another source of Marxism – romanticism – is not included in this analysis)[3]. According to his own words, Karl Marx aimed "to lay bare the economic law of motion of society". For Marx, market economy was disordered and irrational and should have been replaced by the "scientific planning" of production and consumption. The existing capitalist institutions should have been smashed up and replaced by new ones, rationally designed on the basis of scientific knowledge.

Instead of rational institutions, the spontaneous dynamics of the Communist system gave rise to institutions that, in their irrationality,

---

3 Kováč, L., "Natural history of Communism", I. II. *Central Europ. Polit. Sci. Rev.*, 3, 2001, pp. 74-164. http://www.fns.uniba.sk/~kbi/kovlab

had no precedence in history. Instead of a social system ensuring justice and happiness for all, a criminal political system arose – it caused immense human suffering and 100 million people perished. And yet, Communism was not useless providing that we consider it *a posteriori* as a huge social experiment and that we analyse its results and draw appropriate conclusions from them.

It does not seem that this has been undertaken so far. Four features, each of which is rather disquieting, characterise global reactions to Communism:

(1) As a backlash reaction on the failure of planned economy, an extreme view asserting omnipotence of the market is being preached and/or enforced. Faith in the moralising capacity of the market has become almost the new credo for economic reformers in post-Communist countries. Any attempts to control the market and to minimise its imperfections are being harshly criticised – a comeback by Hegel, no longer in Marxist disguise, but in a new form of marker deification, as if the market were an expression of Hegel's Absolute Reason, Spirit of History. In addition, as a reaction to the Communist misuse of power, a new naïve idea is spreading according to which a State should be just a self-service facility sustained by taxes paid by a sovereign individual for guaranteeing his/her safety and creating conditions for his/her full "self-realisation". This exaggerated, and often dilettante would-be liberalism has been called paraliberalism. George Soros has named it "market fundamentalism". By its ignorance of what has already been scientifically established, paraliberalism represents a cultural regress[4].

(2) The collapse of the Marxist "scientific world-view" and refutation of belief in the unlimited capacities of individual reason is sometimes erroneously conceived as a total failure of rationalism and as the end of the era of reason in general. Marxism should be replaced by

---

4 Kcváč, L., 2001, *op.cit.*

postmodernism. The objectivity of scientific knowledge, however restricted, is being questioned: science is conceived as one of several "meganarrations" equivalent to myth, religion, and art. All cultures of the world are being considered as equally valuable, disregarding the number of evolutionary trials they have accomplished, the amount of evolutionary knowledge they have built into their structure, and their tolerance of other cultures.

(3) In the Communism era, Marxism, in its primitive Leninist, Stalinist and Maoist version, became the world-view and ideology of quite a sizeable portion of the population in many countries of the world. It had the capacity effectively to suppress, or even replace, other views, such as chiliastic Christianity or Islam. After the disappointment of "bankrupting" Communism, Marxism has induced a return to original views, and also prepared the ground for the easy rooting and running wild of their fanatic, fundamentalist versions.

(4) Marxist utopia is surviving in academic enclaves in the West, backed by the conviction that Communism failed because it had been implemented in backward countries, therefore the failure cannot be taken as a proof that, in principle, social utopia cannot be accomplished. Even more blatant is tolerance of the political ambitions of Communist parties in post-Communist countries.

In order to make adequate reactions on Communism and to prevent repetition of similar horrific social experiments – perhaps in the form of authoritative regimes deriving their legitimacy not from a "scientific" reason but from a "revealed" reason – superficial, belittling or ignorant views on Communism should be globally supplanted by the competent knowledge of its theoretical basis and deeds. This knowledge provides a weighty corroboration of what contemporary biology says: humans are fearful, hypersocial (and group-confined), hypermoticnal and mythophilic animals[5].

---

5 Kováč, L., 2001, *op.cit.*

⬤⬤⬤ Principle of minimal Prejudice: Reasonableness and Limits of Democracy

Human mythophilia is firmly rooted in one of the strongest human needs: the need to understand the world in which one lives. This is apparently as strong as the need eat and for sexual gratification. The term "understanding" is not equal to the term "knowing". Understanding must be simple, consistent and all encompassing. With restricted knowledge available to an individual, what else can his/her total understanding be if not a myth?

Human capacity of perception, affection and reasoning is a product of biological evolution and is species-specific. All the evolutionary constraints of the human mind confine humans to the world of medium dimension and low complexity. The world outside is counter-intuitive and separated by barriers which have been called "Kant's barriers[6]". The stretch of cultural evolution, an evolutionary "twinkle" between contemporary humans and their primate ancestors, has been too short to have changed the structure of the ancestral human mind to any substantial extent. As Susumo Ohno judiciously remarked, the inner intelligence of human individuals might be quite meagre, just a little higher than that of a chimpanzee[7]. Cultural evolution has not been improving human Reason, but, thanks to emergence of articulated speech, it has been quickly generating new and new pieces of knowledge. Progress of humankind is not progress of individual human Reason, but progress of knowledge and growth of collective intelligence. New pieces of knowledge, temporarily stored in various memory devices, have been built into material artefacts, tools and machines, but also into "social artifacts" – institutions. The first article of European cultural creed – the sacred belief in supremacy of the

131

---

6 Kováč, L., 2001, *op.cit.*
7 Ohno, S., "Genes and the inner conflicts of being man", *Persp. Biol. Med.*, 22, 1978, pp. 3-9.

sovereign, rational, knowledgeable, free and responsible human individual – may hardly rely on getting a backing from biology.

In the course of the growth of human knowledge, a principle emerged which human reason should strictly observe: we should not claim that we know more than we know. It is the principle of minimal prejudice. It has been anticipated by many philosophers and scientists in such ideas as Occam's razor, economy of thought, parsimony, and so on. It may be called Jaynes' principle, according to the physicist who first gave it a precise mathematical formulation[8]. According to E. T. Jaynes, if one has an incomplete knowledge of the subject, the minimally prejudiced assignment of probabilities is that which maximises Shannon's entropy, subject to the given information. The corollary of his argument has been the demonstration that the laws of thermodynamics can be derived as consequences of the principle. It may not be too exaggerated to expect that foundations of some other disciplines of science may also be derived from Jaynes' principle. It seems that its application may provide an innovative view on the nature of democracy.

Genesis of democracy, like genesis of many other social phenomena, has a hybrid character: democracy is partly a product of historical trials and failures, and partly a product of reflections and conscious intentions aimed at improving forms of social life. Spontaneity of the origins and evolution of democracy may well be inferred from the fact that British democracy lacks a written constitution. On the other hand, the founding fathers of the USA deliberately formulated the principles on which the American Constitution should have been grounded. The ingenuity of the American Constitution and, in turn, of the political formation built on it, is due to the fact that the founding fathers observed – obviously, unknowingly – the principle of minimal

---

8 JAYNES, E. T., « Information theory and statistical mechanics », *Phys. Rev.*, 106, 1957, pp. 620-630.

prejudice: they did not claim anything more about human nature and society than what had been known in their time[9]. This virtue gave American democracy its stability and robustness, as well as its capacity of continual improvement and perfection. By this, it also created preconditions for institutional support for the permanent growth of knowledge and thus for the unique success of American science.

Yet, in the present state of American democracy, and of democracy in the world in general, some disquieting questions may no longer be ignored. The event of September 11, 2001, and even more so its aftermath, make these questions most urgent:

(1) Can spontaneous improvement of the system still keep pace with the enormous speed of accumulation of new knowledge, with the dizzy progress of science, technoscience and technology? Have not some founding principles been surmounted and are no longer adequate? Are not some tenets of the American Constitution, correctly reflecting the achieved state of knowledge at the end of the 18th century, outdated nowadays, retarding elements, and impeding due institutional and ideological adjustments?

(2) Notwithstanding the assumption that democracy, at least American democracy, was constituted on the principle of minimal prejudice, are not its very foundations built in powerful ancient myths and had not the very myths given democracy its genuine stability? Has not the Judeo-Christian moral, with its awe of transcendent authority and its fear of God's punishment and eternal damnation, been the essential glue holding together all the components of the edifice of democracy? Is not the erosion of traditional values under the aggressive impacts of science and technology concomitantly the erosion of the very foundation of democratic order[10]?

---

9 KOVÁČ, L., 2001, *op.cit.*
10 KOVÁČ, L., 2001, *op.cit.*

And, ultimately: (3) Should progression of science and technology be slowed down so that social institutions are able, by the rate of their spontaneous changes, to adapt and adjust? It has already been pointed out that humanity has attained a most precarious evolutionary stage: we can do too much in a situation when we understand too little[11]. Yet, the third question may be quite illusory: Any means of how to slow down or stop progression of science and technology – apparently still accelerating and undergoing exponential growth – are unimaginable. Nevertheless, the question should be asked as the future of humankind will be conditioned by the answer.

∷ ● ● From fumbling Democracy to experimental Noocracy

Marxist "scientific management of society" should have been executed by the proletariat, uneducated but endowed with some mystic class instinct enabling it to see and act in the right way. The attempt to install the utopia ended in disaster. Implementation of other utopias would hardly end any better. The scientific dictatorship, devised by August Comte, with a ruling group of bankers and the best scientist in the leading role of the Grand Priest, would probably have been no less irrational and cruel than Lenin's dictatorship of the proletariat. The same may apply to a political and social system designed and ruled by the monopoly of Sigmund Freud's doctrine.

If sustainable social systems, constructed by design from conscious human reason, are not feasible, the method of trial and error seems to be the only workable system. It has been proved to work in biological evolution with such remarkable results that, until Darwin unravelled the principle of natural selection, divine creation had been the only plausible explanation of such feats. It has proved competent in

---

11 Kováč, L., 2001, op.cit.

biological evolution, giving birth to democracy. By trial and error, democracy has been able to improve and adapt to the environment in which it had to sustain itself. What should be undertaken now when the changes to the environment, caused by science and technology, are happening too quickly?

There may be one single possibility of mastering the situation: to speed up the frequency of trials and failures in democracy in the same way as science did in the cultural evolution – by conscious, premeditated, systematic and institutionalised experimentation. Had Communism been deliberately intended to be an experiment, it would have been tried on much smaller scales, in distinct variants, and it would have not been paid for by hecatombs of human lives. From deliberated, scientifically founded experimentation (i.e. randomised, double-blind, placebo-controlled trials) a new method and form of politics can be devised. If Plato called his idea of government sophocracy, the political system with institutionalised science-based social experimentation may be called noocracy. Noocracy would not be the Platonian rule of kings-philosophers. It would not be a reign of science, nor a reign of scientists. Power would continue to be in the hands of political élites, gained and maintained by competition, but of élites professionally trained, exploiting analyses, prognoses and proposals of countless advisory groups of experts from all branches of science, and organising field experiments. The traditional division of roles between politicians and intellectuals would be maintained[12], but the importance of intellectuals as generators of testable ideas would increase appreciably. Experimentation should obviously be applied to the search for new forms of voting systems, participatory democracy (*e.g.* the obligation of deputies to meet criteria of economic and juristic literacy), fiscal policies, wealth redistribution, and education and art supports (with a clear-cut distinction between commodities and

135

---

12 Kováč, L., 2001, *op.cit.*

public goods). Experimental economy may also be instituted in noocracy. (Experimental economy should not be confused with experimental economics which already thrives as a branch of economics and which comprises experimentation at the laboratory scale.)

The case in point may be the recent controversy concerning the use of genetically modified organisms (GMOs). In full-fledged noocracy, GMOs would be tried in one or several regions or countries and scientifically monitored by all, under the auspices of a central governing body. Costs and benefits would eventually be shared equally by all, and administered and imposed by the central, democratically constituted authority. The widely disputed precautionary principle would be in operation without, at the same time, slowing down or hindering the application of scientific inventions.

Conclusion: A "Star" Opportunity for the European Union

There are exceptional instances in history, which – by paraphrasing Stefan Zweig – may be called the "star opportunities": they are short-lived, yet the future long-term evolutionary trajectory of a society may be decided in just such an instant. The European Union now finds itself in such a star situation. It can turn into a considerably centralised, rigid, bureaucratic superstate, but it can also become a political system consistently built on the idea of evolutionary rationalism. In this way, it would serve as an unprecedented research institution of political and social experimentation – a pioneer of noocracy. This presupposes preservation of the existing plurality of institutional structures as represented by particular states and nations, and even its expansion by strengthening autonomy and authority of intrastate and interstate regions. Every such relatively autonomous political unit would represent a single political laboratory. In each of these, hypotheses, corresponding to available scientific knowledge, would be subjected to testing. The lesson, drawn from falsification of a hypothesis tested in one laboratory, would serve as a warning to all: there is no need of trying that way any further. The results of successful experiments would be disseminated all over the world.

The result of Communism, which had not been intended as an experiment but could be analysed as such, falsified some fundamental hypotheses on human nature and social dynamics. If properly evaluated and exploited it may turn into a unique asset of the new European Union Member States. Instead of being considered as poor relatives of the core members, their experience with the failed attempt of a "scientifically managed society" may become their substantial contribution to the joint cultural treasury.

Renunciation of traditional European rationalism does not mean its repudiation and condemnation. After its birth in Ancient Greece, rationalism has become the major evolutionary impetus, not only for Europe but for the whole world. As regards the achievements of science, it needs rectification and transformation. Abandoning the illusion of the boundless potency of human reason and promoting new, evolutionary rationalism, Europe – polymorphous and experimenting – may resume its leading role in promoting the advancement of global civilisation.

● ● ● Acknowledgment

This study was supported in part by Howard Hughes Medical Institute Research Grant No. 55000327.

SCIENCE WITHOUT A CONSCIENCE

Sophie Bessis

Sophie Bessis has been professor of development and international co-operation at the University of Paris I-Sorbonne for the past eight years. Her work focuses on world food exchanges and the political economy of development in Africa and the Middle East. During her rich and varied career, she has taught history at the University of Yaoundé (Cameroon), and was a professional journalist for nearly 15 years. She has also worked for UNICEF, UNESCO and other international organisations. Her published books in French include *The food weapon*, *The final frontier*, *Hunger in the world* and *The children of the Sahel*.

Should scientists be personally concerned about the consequences of their discoveries, or are they exempted from all responsibility for any use to which they may be put? In other words, does science have to render account to society, or should it remain untrammelled by any democratic control? Science, of course, is not the only human activity to escape such control but, today, its fields of investigation and its servants' stubborn desire to see their discoveries widely used – regardless of any current or future risk – pose the question with a new urgency.

● ● ● The scientific Exception

For several decades, scientists – those who "know" and, as such, implicitly opposed to the mass that do not – have been depicted as impartial researchers spurred on by the sole quest for knowledge, pursuing this noble aim in the solitude of their laboratories, sheltered from the collective prejudices and fears of normal mortals. Science erjoys a sort of weightlessness detaching it from any of the geographic, cultural, temporal or historical contingences which generally mark intellectual production. It is sovereignly independent and, to remain thus, must stay shielded from the discussions of the Polis. This pious image, zealously maintained by the scientific community, but which is fortunately wearing thin at the edges, makes so-called scientific disciplines the only products of the human intellect that claim the right to be totally independent of the historical, sociological or economic contexts in which they carry out their labours.

Science and Progress

This exorbitant privilege, by which science sets itself apart from any other intellectual or technical activity, and which is accepted by society at large, is explained by a conviction that it represents one of the pillars of modern-day thought, and that every scientific discovery marks another step in the forward march of progress that is specific to the human race. Starting from the point in time – between the 16th and 18th centuries – when religious authorities in Europe began to lose their magisterium and power, science and progress became synonymous, the former as a vital partner to the latter that can no longer be conceived without it. Since then, any critical question directed at science and its objectives is presented as an unacceptable rejection of progress and a unanimously stigmatised desire to have humanity move backwards. The only enemies of science – the perfect outworking of reason, seeking as its sole objective to accelerate the inevitable distancing of humanity from its original natural state – are the rearguard of the old order, in thrall to obscurantism, the irrational and magic.

Examples abound of the ease with which scientists dismiss any criticism as unenlightened ignorance. Just as opponents of nuclear energy are suspected of wanting to return to candles and rushlights, people opposed to the use of genetically modified organisms (GMOs) in agriculture until their effects are better known find themselves accused of indifference towards the planet's underdeveloped and starving masses. "The GMO revolution is an essential advance" to save humanity from famine, writes French biochemist Etienne-Emile Baulieu, but "it would appear as difficult to accustom ourselves to GMOs at the beginning of the 21st century as it was to step into trains in the mid-19th, when people feared suffocating in tunnels …[1]"

At the international Earth Summit organised by the United Nations in June 1992 in Rio de Janeiro, which opened one of the first major world-level debates on the ultimate objectives of growth and the excesses of productivism, 260 scientists, including 52 Nobel prize-winners, signed the "Heidelberg Appeal". This veritable declaration of war against what we would call the precautionary principle today, declared, among other things: "We, the undersigned, members of the international scientific and intellectual community […] are worried at the dawn of the 21st century, by the emergence of an irrational ideology which is opposed to scientific and industrial progress and impedes economic and social development. […] We forewarn the authorities in charge of our planet's destiny against decisions which are supported by pseudoscientific arguments or false and non-relevant data. […] The greatest evils which stalk our Earth are ignorance and oppression, and not science, technology, and industry whose instruments, when adequately managed, are indispensable tools of a future shaped by humanity, by itself and for itself, overcoming major problems like overpopulation, starvation and worldwide diseases." In short: the concern for the consequences to the planet of the headlong rush into a technologist future that characterises contemporary civilisation is tantamount to opposing progress.

---

1 In *Le Monde*, 22 October 2003

With such expressions, scientific discourse postulates that we must believe in the essentially progressive character of science, and that such belief is not negotiable. Indeed, this postulate itself is closed to discussion and anyone having the temerity to question it is an adversary of progress, and hence of humanity itself. The rightful place for discussion, if any, is within the Academy, among specialists recognised by their peers. Such is the dominant discourse by which contemporary scientists protect themselves from the slightest questioning by society.

Scientific Position and religious Status

Whatever the tenants of scientific rationalism might say, this position bears more than a faint resemblance to that of the religious authorities. It reveals a curious paradox: science, which for centuries has fought to free itself from the dictates of church, synagogue or mosque, appears to wish to legitimate its enterprises by resorting to the same processes as applied by those which, for many years, it denounced as the worst enemies of reason. Our faith in science must be unquestioning, like our forefathers' faith in God. Any other attitude is sacrilege. Surrounded with this aura, science is understandably unwilling to bend its proud head to enter into the gate of the Polis.

So, has scientific discourse replaced the word of religion ever since western societies opted to live under the sceptre of reason? A daring hypothesis perhaps, but we cannot deny the similarity of the way in which science has arrogated to itself, in place of religion, a monopoly on the production of truth. Indeed, on this central point of its argument, the scientific community cultivates a surprising historical amnesia, forgetting that certain of its former truths are today considered as errors – at best naïve, at worst the source of tragic consequences. With God out of the picture, only science can tell us the truth today.

This probably explains why all contemporary endeavours to control society have sought to drape themselves with the mantle of science in order to impose their conception of the world, or – more modestly – of the organisation of society. Let us remind ourselves how, for several decades, socialism was "scientific", as scientists were anxious to prove. On another level, economics is constantly being passed off as an exact science, with "laws" that, all of a sudden, are written in stone and binding on everyone, without discussion.

Science in History

However much it might comfort and console the research community, we have to ask whether this position is tenable. Is it not time to put an end to the "scientific exception" and to bring science into the democratic debate? Can we still assert that every discovery advances the well-being of humanity? May one not wish that society could have a word to say about scientific choices and their implications? We know the comment rightly or wrongly attributed to Einstein, in despair at having contributed – through his genius – to producing the atom bomb: one day the scientist exclaimed that he would have done better to take over his father's cobbler's shop. More seriously, it is perhaps time to remind science that none of its choices or its discourses is neutral, and it can no longer use its lack of responsibility as an excuse for not setting itself limits.

This is all the more vital because, contrary to the imagery evoked at the beginning of this text, a scientist in no longer the solitary man of knowledge, independent of churches and princes, poor perhaps, but free. It is stating the obvious to remind ourselves of the extent to which science, undertaken in teams in extraordinarily sophisticated laboratories, is a highly expensive undertaking, devouring ever greater sums of money, most of it from private companies with the balance coming from specifically targeted public programmes.

However much he may cling to his icon, our researcher is increasingly a "science worker" whose research programme is set by his employers' objectives and strategies.

With the current enthusiasm for life sciences, breaches are indeed beginning to appear in science's discourse on itself, structured for too long by the desire to remain outside history and to free itself from the control of the Polis. But this obstinacy forgets two – among many other – lessons of history: no intellectual production can be read outside its context, and science can also go wrong where its context incites it to.

● Physical Anthropology – a Precedent

To convince ourselves of this, let us dwell for a moment on the discourse of 19th century physical anthropology which, as we know, provided scientific underpinning to the colonial enterprise of the imperialist powers. This they used and abused widely to legitimate their conquests and domination. Out of this new discipline was born modern racism, this scientific theorisation of European supremacy. According to this doctrine, racially superior "Caucasians" or "Aryans" are designated by science to reign as masters over a humanity made up of a series of groups ranked hierarchically by the distance separating them from the chosen race. In his habitual fashion, Renan summarises in luminous language the scientific "advances" of his day: "Nature has created a race of workers. These are the Chinese, with dexterity of hand and almost no sense of honour […]; a race of workers of the soil. These are the Negroes [...]; a race of masters and soldiers, the Europeans[2]."

Even among Blacks, the closest to the animal state, certain groups are more human than others, being less "Negroid" in features and colour.

---

2 RENAN, E., *Œuvres complètes*, 1947.

Here too, science undertook to "prove" that brain size is directly proportional to the lightness of the skin colour. The "Hamites" of Africa's Great Lakes – an artificial racial category if ever there was one – saw themselves designated as the whitest of Negroes, with the privileges implied by such status. Superiority finds a simple explanation in physical features, from which it derives the genius – scientific, technical, cultural and political – specific to the white race. Cohorts of scientists worked throughout the 19th and into the 20th century to refine these discoveries, using an array of scientific tools, such as cranometrics or anthropometrics, to measure the degree of primitivism of subjected populations by their "facial angle", "nasal index" or "cephaloid index".

Going no further than France, science's caution towards the expression of such racism counted for nought. In 1877, the Parisian *Jardin Zoologique d'Acclimatation* inaugurated its exhibitions of natives imported from exotic regions and displayed like animals. Until 1886, the Paris Anthropological Society supported these exhibitions, appointing a commission chaired by surgeon and anthropologist Paul Broca to "examine carefully the natives camped at the gates of Paris[3]". At the 1889 Universal Exhibition, Doctor Letourneau, director of the cabinet of anthropology, commented on the crania of various races on display: "anatomical anthropology teaches us that [...] the Negroid races are at the bottom of the ladder and white races at the top[4]."

We know the damage caused by such madness, both in the colonies and in Europe itself. Closer to home, the recurrent debate, which opposes partisans of the genetic and environmental schools of human behaviourism, is itself highly coloured by the political and economic context which, in turn, explains the alternating fortunes of these two disciplines. As far back as 1949, American Nicholas Pastore,

---

3 Cited by MANCERON, G., *Marianne et les colonies*, Paris, La Découverte, 2002.
4 MANCERON, G., 2002, *op.cit.*

examining in *The Nature-Nurture Controversy*[5] the correlation between the political commitment of certain scientists and their position in the heredity-environment debate, concluded that their scientific opinion was conditioned decisively by their position on the political chequerboard.

Bringing Science back into the Polis

Without wishing to insult the scientific community or deride the innumerable contributions of its discoveries, these summary reminders of certain historical facts are intended first of all to reaffirm that truth is always relative. But not only that – is it not time to remind the West that the demiurges populating its mythology are not all benefactors of humanity, as their praise-singers would have us believe? For every one Prometheus bringing fire to mankind, how many sorcerers' apprentices have produced Frankensteins or Golems? The dominant scientific discourse remains the offspring of the civilisation producing it, with its desire to rob the gods of their wisdom and to replace them, whatever the consequences, along with the monotheistic belief in the absolute nature of truth. In order to bring science into the Polis, it is high time therefore to loosen the ties that bind science, progress and truth. This will be no easy task as long as this trinity structures modern thought. But it is also the precondition for finally breaching the sacred ramparts of the word of science, allowing it to be discussed like any other word.

Time is short. This sacrality, which historically has served other designs than those of science alone, as we have seen above, has today become the tool of enormous economic and financial forces. The science-progress-truth triptych is in ever-greater danger of being pressed into the service of logics of profit, with even fewer limits than

5 PASTORE, N., *The Nature-Nurture Controversy*, New York, King's Crown Press, 1949.

those set by the heralds of the Promethean myth. A handful of doubtful experiments, like the famous "Terminator" gene that can sterilise seed, lead us to fear the mass appropriation by the commercial sector of the discoveries of the new science workers. As in other domains, only democratic control can put a stop to potential misuses. History teaches us of the damage done when the "scientific exception" becomes the servant of domination. Equally tragic harm could result if science today becomes the servant of a handful of business enterprises under pressure to find new sources of profit. It is to guard their independence and not to clip their legitimate margin of autonomy that scientists need to accept they must respond to citizens' questions. Much of the scientific community has already understood this, and is debating intensely the need for science and society to meet. Such controversy is cause for optimism, because it is scientist-citizens who will enable society to say no to sorcerers' apprentices.

SCIENCE, ETHICS AND DANGER

Lewis Wolpert

Lewis Wolpert is a professor at the department of anatomy and development biology at London University (UK). After originally training as a civil engineer in South Africa, he entered the world of cell biology in 1955. He became a fellow of the Royal Society (UK) in 1980 and a Commander of the Order of the British Empire (CBE) in 1990. He has presented a number of popular science programmes. In addition to writing a newspaper column, he has published several books, the most recent of which was *The unnatural nature of science*.

re scientists for the applications of science? In a recent issue of the journal *Science*, the 1995 Nobel Peace Prize laureate, Sir Joseph Rotblat, proposes a Hippocratic oath for scientists. He is strongly opposed to the idea that science is neutral and that scientists are not to be blamed for its misapplication. Therefore he proposes an oath, or pledge, initiated by the Pugwash Group in the United States. "I promise to work for a better world, where science and technology are used in socially responsible ways. I will not use my education for any purpose intended to harm human beings or the environment. Throughout my career, I will consider the ethical implications of my work before I take action. While the demands placed upon me might be great, I sign this declaration because I recognise that individual responsibility is the first step on the path to peace."

These are indeed noble aims to which all citizens should wish to subscribe, but the oath does present some severe difficulties in relation to science. Rotblat does not want to distinguish between scientific knowledge and its applications, technology, but ignores the fact that the very nature of science is that it is not possible to predict what will be discovered or how these discoveries could be applied. Cloning provides a good example. The original studies related to cloning were largely the work of biologists in the 1960s. They were studying howfrog embryos develop and wanted to find out if genes which are located in the cell nucleus were lost or permanently turned off as the embryo developed. It was incidental to the experiment that the frog that developed was a clone of the animal from which the nucleus was obtained. The history of science is filled with such examples. The poet Paul Valery's remark that "We enter the future backwards" is very apposite in relation to the possibleapplications of science. Scientists cannot easily predict the social and technological implications of their current research. It was originally argued that radio waves would have no practical applications, and Lord Rutherford said that the application of atomic energy was moonshine. Similarly, there was no way that those investigating the ability of certain bacteria to resist infection by viruses would lead to the discovery of restriction enzymes, an indispensable tool for cutting up DNA, the genetic material which is fundamental to genetic engineering.

Part of the problem is the all too common conflation of science and technology. The distinction between science and technology, between knowledge and understanding on the one hand, and the application of that knowledge to making something, or to use it in some practical way, is fundamental. Science produces ideas about how the world works, whereas the ideas in technology result in usable objects. Technology is much older than anything one could regard as science and unaided by any science; technology gave rise to the crafts of early humans, such as agriculture and metalworking. Science made virtually no contribution to technology until the 19th century. And even the great triumphs of engineering, like the steam engine and Renaissance cathedrals, were built with virtually no impact from science. It was imaginative trial and error. It is technology that carries with it ethical issues, from motor cars to cloning a human.

By contrast, reliable scientific knowledge is value-free and has no moral or ethical value. Science tells us how the world is – that we are not at the centre of the universe is neither good nor bad, nor is the possibility that genes can influence our intelligence or our behaviour. Dangers and ethical issues only occur when science is applied as technology. However, ethical issues can arise in actually carrying out the scientific research, such as doing experiments on humans or animals, as well as issues related to safety. In this respect, scientists have very similar ethical problems to those of all citizens.

The social and ethical obligations that scientists have as distinct from those responsibilities they share with all citizens, such as supporting a democratic society and taking due care of the rights of others, comes from them having access to specialised knowledge of how the world works, which is not readily accessible to others. Their obligation is to both make public any social implications of their work and its technological applications, and to give some assessment of its reliability. In most areas of science it matters little to the public whether a particular theory is right or wrong, but in some areas – such as human and plant genetics – it matters a great deal. Whatever new technology is introduced, it is not for the scientists to make the moral or ethical decisions. They have neither special rights nor skills in areas involving moral or ethical issues. There is, in fact, a grave danger in asking scientists to be more socially responsible if that means that they have the right and power to take such decisions on their own. Moreover, scientists rarely have power in relation to applications of science; this rests with those with the money – industry and government. The way scientific knowledge is used raises ethical issues for everyone involved, not just scientists.

It is not easy to find examples of scientists as a group behaving immorally or in a dangerous manner – BSE is not an example – but the classic was the eugenics movement. The scientific assumptions behind this proposal are crucial; the assumption is that most desirable and undesirable human attributes are inherited. Not only was talent perceived of as being inherited but so too were pauperism, insanity, and any kind of so-called feeble-mindedness. Thos involved completely failed to give an assessment of the reliability of their ideas. On the contrary, and even more blameworthy, their conclusions seem to have been driven by what they saw as the desirable social implications. By contrast, in relation to the building of the atomic bomb, the Allied scientists behaved morally and fulfilled their social obligations by informing their governments about the implications of atomic theory. The decision to build the bomb was taken by politicians, not scientists, and it was an enormous engineering enterprise. Had they decided not to participate in building an atomic weapon, that could have lead to the war being lost. Should scientists on their own ever be entitled to make such decisions?

Genetics seems to raise many ethical issues. Mary Shelley would be both proud and shocked. Her creation of a scientist, Frankenstein, creating and meddling with human life has become the most potent symbol of modern science. Her brilliant fantasy has become so distorted that even those who are normally quite sensible lose all sense when the idea of cloning humans appears before them. The image of Frankenstein has been turned by the media into genetic pornography whose real aim is to titillate, excite and frighten. The bio-moralists are triumphant with the aid of genetic pornography to titillate and frighten, as purveyed by the media. Cloning is the most obvious example of this.

Ironically, the real clone of sheep has been the media blindly and unthinkingly following each other – how embarrassed Dolly should have been. The moral masturbators have been out in force telling us of the horrors of cloning. Jeremy Rifkin in the USA demanded a world-wide ban and suggests that it should carry a penalty "on a par with rape, child abuse and murder". Many others, national leaders included have joined in that chorus of horror. But what horrors? What ethical issues? In all the righteous indignation I have not found a single relevant new ethical issue spelled out.

The really important issue is how the child will be cared for. Given the terrible things that humans are reported to do each other and even to children, cloning should take a very low priority in our list of anxieties. Or perhaps it is a way of displacing our real problems with unreal ones. Having a child raises real ethical problems as it is parents who play God, not scientists. Here lies a bitter irony. A parent's relation to a child is infinitely more God-like than anything that scientists may discover. Parents hold tremendous power over young children, but they do not always exercise it to the child's benefit.

This aspect of parental care is very relevant to the so-called ethical issues raised by designer babies. I cannot see why parents should not be able to choose a child to have genes that promote good health and avoid those that lead to disease. And why should they not seek intelligence or blue eyes although, for a long time, the former, involving thousands of genes is most impractical. Much more serious and neglected is the right of couples who are child abusers and drug addicts and with serious genetic illnesses to have children. There are tens of thousands of children referred to social services each year and about 10% of all children suffer form some sort of abuse – physical, emotional, psychological. These are the real issues in relation to reproduction.

Cloning of a human raises no new ethical issues and should be opposed on the grounds that there is a risk of the child developing abnormally. The use of stem cells and therapeutic cloning to make stem cells that could provide tissues to replace damaged organs without the increased risk of immune rejection raises no such problems. No politician has publicly pointed out or even understood that the so-called ethical issues involved in therapeutic cloning are indistinguishable from those that are involved in *in vitro* fertilisation – IVF. One could even argue that IVF is less ethical than therapeutic cloning. But no reasonable person could possibly want to ban IVF that has helped so many infertile couples. Where are the politicians who will stand up and say this?

And what are the objections to making embryonic stem cells. The fertilised egg is not a human being. Would one not rather accept a thousand abortions and the destruction of all unwanted frozen embryos than a single unwanted child who will be neglected or abused. I take the same view as regards severely crippling and painful genetic diseases. On what ground should parents be allowed to have a severely disabled child when this could be relatively easily prevented by prenatal diagnosis? It is nothing to do with consumerism but the interests and rights of the child.

So what dangers does genetics pose? Bioethics is a growth industry but one should regard the field with caution as the bioethicists have a vested interest in finding difficulties. Anxieties about designer babies are at present premature as this is far too risky, and we may have, in the first instance, to accept what Ronald Dworkin has called "procreative autonomy", a couple's right "to control their own role in procreation unless the state has a compelling reason for denying them that control". Gene therapy, introducing genes to cure a genetic disease like cystic fibrosis, carries risks – as do all new medical treatments. There may well be problems with insurance and testing but are these any different from those related to someone suspected of having AIDS? One must wonder why the bio-moralists do not devote their attention to other technical advances like that convenient form of transport which claims over 50,000 killed or seriously injured each year. Could it be that in this case they themselves would be inconvenienced?

But there are surveys that show some distrust of scientists, particularly those in government and industry. This probably relates to BSE and GM foods, so one must ask how in fact this affects people's behaviour. I need to be persuaded that many of those who have this claimed distrust would refuse, if ill, to take a drug that had been made from a genetically modified plant, or would reject a tomato thus modified that is was both cheap and would help prevent heart disease. Who refuses insulin or growth hormone because it is made in genetically modified bacteria? It is easy to be negative about science if it does not affect your actions.

Genetically modified foods have raised extensive public concerns, and there seems to be no alternative but to rely on regulatory bodies to assess their safety as they do with other foods – similar considerations apply to the release of genetically modified organisms. Genetic engineering requires considerable scientific and technical knowledge and even more important – money – which scientists, in general, do not have. Indeed, for the public sector, the applications of genetics and molecular biology can open up difficult choices because such applications are expensive. New medical treatments, requiring complex technology, cannot be given to everyone. There has to be some principle of rationing and this really does pose serious moral and ethical dilemmas much more worthy of consideration than the dangers posed by genetic engineering.

Are there areas of research that are so socially sensitive that research into them should be avoided, or even proscribed? One possible area is that of the genetic basis of intelligence and, in particular, the possible link between race and intelligence. So are there then, as the literary critic George Steiner has argued, "certain orders of truth which would infect the marrow of politics and would poison beyond all cure the already tense relations between social classes and these communities"? Also, it has even been suggested by Kitcher that work on the human genome should stop as it could provide evidence for racial differences which show that one group of people is inferior in certain skills. He is sure that such a finding would further disadvantage them; he may be right, but he does not even consider a society which would positively compensate them as is often done for those with genetic illnesses. In short, are there doors immediately in front of current research which should be marked "Too dangerous to open[1]"? I realise the dangers but I cherish the openness of scientific investigation too much to put up such a notice. I stand by the distinction between knowledge of the world and how it is used. So I must say "no" to Steiner's question, provided of course that scientists fulfil their social obligations. The main reason is that the better understanding we have of the world, the better chance we have of making a just society, and the better chance we have of improving living conditions. One should not abandon the possibility of doing good by applying some scientific idea because one can also use it to do bad. All techniques can be abused and there is no knowledge or information that is not susceptible to manipulation for evil purposes. I can do terrible damage to someone when I use my glasses as a weapon. Once one begins to censor the acquisition of reliable scientific knowledge, one is on the most slippery of slippery slopes.

To those who doubt whether the public or politicians are capable to taking the correct decisions in relation to science and its applications, I strongly commend the advice of Thomas Jefferson: "I know no safe depository of the ultimate powers of the society but the people themselves, and if we think them not enlightened enough to exercise that control with a wholesome discretion, the remedy is not to take it from them, but to inform their direction." But how does one ensure that the public are involved in decision-making? How can we ensure that scientists, doctors, engineers, bioethicists and other experts, who must be involved, do not appropriate decision-making for themselves. How do we ensure that scientists take on the social obligation of making the implications of their work public? We have to rely on the many institutions of a democratic society: parliament, a free and vigorous press, affected groups, and the scientists themselves.

---

1 Wolpert, L., "Is science dangerous?", *J.Mol.Biol.*, 319, 2002, pp. 969-972.

WISH FULFILMENT AND ITS DISCONTENTS ON THE UNEASY RELATIONSHIP BETWEEN THE LIFE SCIENCES AND THE HUMANITIES[1]

Helga Nowotny

Helga Nowotny has been professor of social studies of science at ETH Zurich since 1996. She is a board member of several scientific and policy-oriented institutions in Europe. Professor Nowotny chairs the European Commission's European Research Advisory Board (EURAB) and is also a member of the European Research Council Expert Group (ERCEG). With law and sociology degrees, she has held teaching and research positions across Europe. She received the *Society for Social Studies of Science's Bernal Prize 2003*, and won the *Arthur Burkhardt Prize for the Promotion of Science 2002*.

157

At this year's Biennale international art exhibition in Venice, Italy, Patricia Piccinini, a Melbourne-based painter and sculptor, presented sculptures of synthetic life forms (Fig. 1). Entitled The Young Family, Leather Landscape and Still Life with Stem Cells, her art crosses the species boundaries between humans and other animals. Treading a fine line between the grotesque and the lifelike, Piccinini's art provokes deeply ambivalent emotions in the viewer. Her monstrous, mutant life forms are never repulsive, but also never take possession of our nurturing instincts. Rather, they convey their difference from and likeness to humans with an impressive dignity of their own. The interface between science and society conjured in these creatures, through the imagination of the artist and the impressive range of media she masters, is a superb

---

1 Article published in *EMBO reports*, VOL 4, N° 10, 2003.

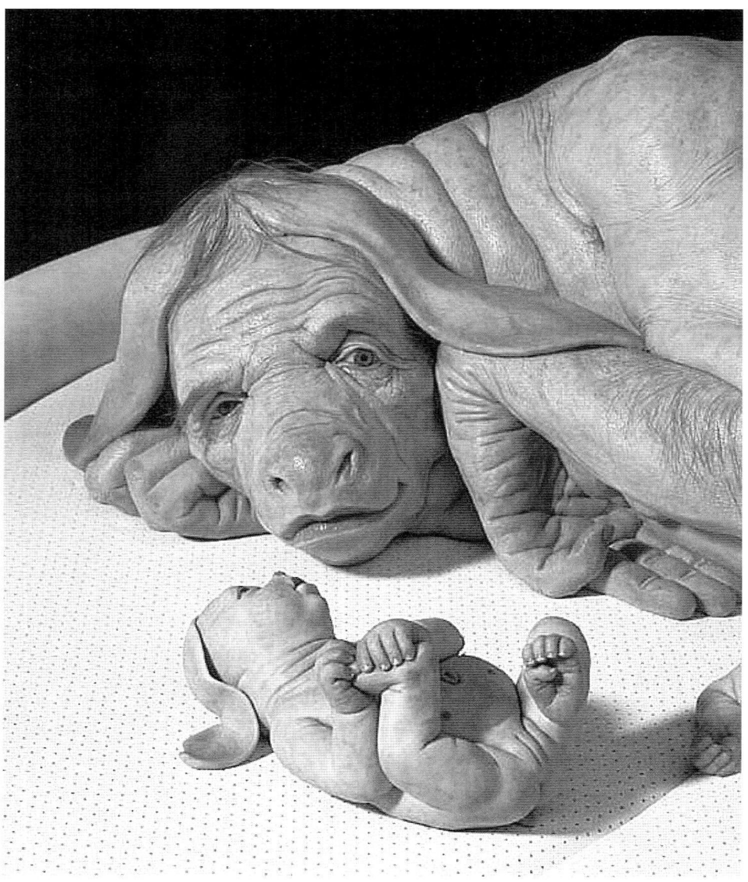

Fig. 1 | *The Young Family,* by Patricia Piccinini (2002). Silicone, acrylic, human hair and leather. Courtesy of the artist and Roslyn Oxley Gallery, Sydney, Australia.

example of how some of the seemingly intractable questions posed by the life sciences can be addressed: by exposing our ambivalent emotions and provoking further reflection and discussion.

Another artist's work at the Biennale also addresses our understanding of humanity. Michal Rovner's projects are carefully choreographed configurations of people who, at first sight, appear to be forming and reforming new patterns. Photographed in black and white, Rovner's art resembles images from the early years of cinematography. The fascination that these ballet-like arrangements evoke comes from the repetition of movements leading to change, and change leading to new repetitions. Images of the same people are projected as walking along horizontal lines, and bear a strong resemblance to patterns of genome sequencingIn this artistic play of scale and size, of form and movement, of repetition, replication and change, the artist creates patterns with human figures, transforming them from human size to the miniature size of bacterial cultures, then transposing them ultimately to the molecular level. The question of what it means to be human is also present here – it demands different, but equally tantalising interpretations.

Both artists address issues that are at the core of genetics and of the research agenda of the life sciences in general. Why is it seemingly easier for the arts than for the humanities – and partly the social sciences – to approach the realm of molecular biology, and especially genetics, with the aim of investigation and understanding? Why is it that the natural sciences and the humanities have increasingly little to say to one another, leaving the growing gap to be filled by tension and mutual distrust? It is indicative of the no-man's land that has emerged from this mutual alienation, that ethics committees now serve as bridges where "normal" exchange between disciplines no longer takes place. Ethics committees and their spokespeople now resemble peacekeeping forces that have been dispatched to patrol the contested ground between enemy territories.

There is no doubt that many new and deep ethical issues are being raised through the advances of biology, calling either for regulation or at least guidelines. With every opportunity that science or the

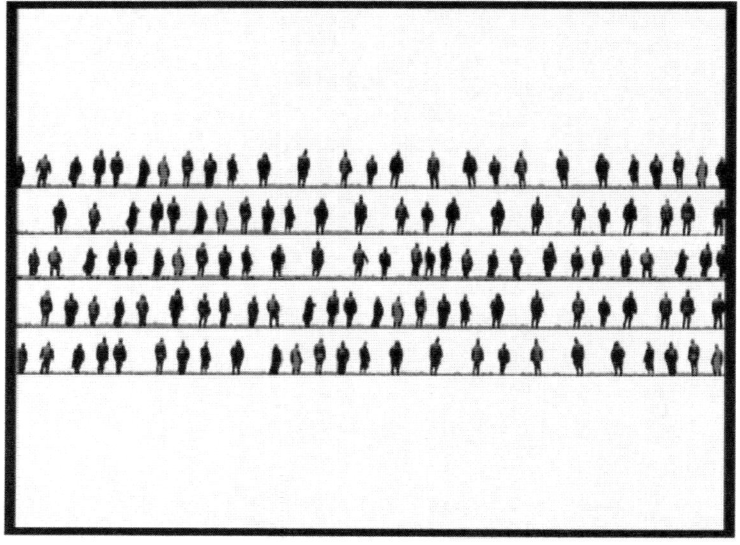

Fig. 2 | *Notes 4*, by Michal Rovner (2002). Pure pigment on archival paper. Edition 1 of 5. Courtesy of the artist and Stephen Friedman Gallery, London, UK.

application of new scientific knowledge and technology creates, new uncertainties arise, just as the fulfilment of a promise may give rise to yet another promise. Modern science and technology – despite, or rather because of their real achievements – have fuelled dreams that resemble a mirage: highly desirable, but ultimately unattainable. Wishes have been fulfilled, but the results differ from the promises that initially fuelled the wish: hence the discontent with wish fulfilment. The responses that the life sciences have given to deeply felt human needs and desires have created new disenchantments with the humanities. Inspired by the old Enlightenment dream of improving our lives, the natural sciences have, perhaps naively or blinded by their own professional ideology, striven towards this goal.

On the uneasy Relationship between the Life Sciences and the Humanities

They have achieved impressive feats without asking themselves too often into which of the several possibilities of human improvement their achievements have fallen. As a consequence, the public has become more and more demanding of the natural sciences and increasingly looks for science and technology to provide solutions for the many problems and ailments they strive to overcome.

Humanists, in turn, see their model of the perfectibility of humankind, or even its desirability, side-stepped. Worse still, their efforts to contribute to a better understanding of the human condition and its improvement through debate and confrontation fall on deaf public ears. Clearly, their answers to the question "what does it mean to be human?" are more diverse and often embedded in cultural meanings that continue to change over time. Now, humanists fear, these answers are increasingly ignored.

This has not always been the case. In fact, the natural sciences and the humanities have seldom been as divided as they are today. The Greek philosophers embodied both the natural philosophy that tried to understand and explain the world in which humans live, and the social philosophy that set out to do the same for human society and its interactions. This unifying approach was again present during the Renaissance and the Enlightenment, when outstanding thinkers such as Leonardo da Vinci, Galileo Galilei, Adam Smith, Voltaire and Johann Wolfgang von Goethe saw no limits to their interests, and the work of these true universalists included the whole range of natural sciences, humanities and the arts. It was only with the universities of the 19th century that the humanities and the natural sciences parted ways and set out to find answers to the same questions, albeit via different paths. Still, the division of scientific labour, institutionalised under the firm control of scientific élites, accorded a highly respectable position to the humanities. It was greatly reinforced through the differentiation of society into separate spheres with different functions: the state, industry, science and culture. Under the widely shared project of modernisation and fuelled by aspirations to modernity, the mechanisms of centralised control rested ultimately with the nation-state. Under the nation-state's tutelage and by forming varying but sustained alliances with it – including the military side – the scientific disciplines flourished: natural sciences as well as humanities.

Although they differed in their ideas about desirable futures or how improvement in material conditions and lives could be reached, this overall division of labour functioned reasonably well, assigning different roles to the natural sciences, the social sciences and the humanities. Of course, controversies repeatedly flared up, beginning with Goethe questioning Newton's theory of colour. Romanticism, the "hidden" side of the Enlightenment, also conjured up alternative visions of what life means, leaving space for emotion and creative inspiration. But the fact was that many of the best natural scientists saw themselves as deeply rooted in a shared, humanistic culture. This was certainly the case for several generations of the most eminent physicists, and ceased only after the Second World War – the end of the heroic age – when professional self-images became increasingly permeated with other, more "strategic" criteria to define science[2].

The humanities therefore felt relatively secure and self-assured in the territories they occupied. They had a major role in shaping national and cultural identities and accordingly were held in high public esteem. Their influence extended to the education system and, together with the legal profession, to the civil service. Culture was another important domain in which the humanities held sway. Of course, there was criticism directed against the acceleration of scientific and technological advance and especially of the all-embracing notion of "progress". But even this critique was mitigated

---

2 Schweber, S., "Physics, community, and the crisis in physical theory", *Physics Today*, 46, 1993, pp. 34-40.

by the shared sense that modernity had its price. Moreover, at least in Europe, the humanities had found a publicly recognised place for making their voice heard: through the social figure of the intellectual, who would speak out in public to warn of a wide range of problems and to provide orientation and critique, while trying to live up to an ideal of how public responsibility on the part of the humanities could be exemplified.

The present situation could not be more different. Everywhere, it seems, the humanities have been relegated to the margins of the universities that they still consider their main intellectual home. Research funding and prestige, media attention and public recognition mixed with resentment have shifted their focus decisively and irreversibly towards the natural sciences. Intellectuals have lost ground, even where they paradoxically were able to hold out longest, in Eastern Europe. Some of the traditional disciplines of the humanities have turned into new specialities, such as cultural studies, that seek outlets for their students closer to the market and the media. It can be argued that there has been a spectacular growth in cultural production that is hardly noticed and certainly not given the attention accorded to innovation and growth in science and technology. The reason for this relative neglect lies in the incorrect assumption that the humanistic connection to the creation of wealth is weak and that cultural production is less costly. A more differentiated assessment reveals, at least partly, the deep involvement of certain forms of cultural production with the production of images and its economy. Cultural products are the symbolic currency in the market of life chances, just as new technological products underlie the hard currency in the markets of industry.

The important point, however, is that the current state of affairs cannot be explained solely by comparing scientific output and results. Nor can the more efficient research methodologies or the technology-driven research agendas of the natural sciences, which make for greater efficiency in research organisation, account for the growing gap.

Computerisation has, with a delay, also entered the humanities. When terrorism came onto the global stage, the knowledge of distant and minor languages and the cultures to which they belong gained spectacular public visibility and importance. Education in music, the arts and literature is still considered desirable in intellectual circles, even if the canon of what constitutes "culture" or what an educated person should know is no longer uncontested. The flirtation with postmodernism that parts of the humanities have engaged in has led to the removal of some false certainties but, more generally, has tested their limits. Faced with the abyss of relativism, the choice for the humanities has been either to vanish within it or to secure new ground. The uneasy relationship with the market has led to a multitude of practical arrangements and accommodations that defy the official rhetoric. To understand the gap we have to go one step further.

As my colleagues and I have pointed out in *Re-Thinking Science*, the broader context of the transformations of the scientific system and the associated controversy has gone largely unexplored[3]. It is increasingly difficult to distinguish between the domains of the state and the market, between culture and mass media, and between what constitutes public and private arenas. Institutional boundaries have become fuzzy and overlapping. Science itself is increasingly challenged by other forms of knowledge production. Paradoxically, the autonomy of science, which was always relative, is no longer guaranteed as its potential guardians – the state, market, media and culture – are no longer recognisable in their former identities and functions. Perhaps equally paradoxically, the advance of science and

---

3 NOWOTNY, H., SCOTT, P. & GIBBONS, M., *Re-Thinking Science Knowledge and the Public in an Age of Uncertainty*, Cambridge, Polity Press, 2001.

technology has enlarged the realm of the "political" sphere, creating the need for an array of regulations and regulatory frameworks that, in a liberal democracy, are preceded by negotiations, mediations, consultations and debate. This is the public space (we call it the *agora*) in which society increasingly "talks back to science" and where scientific expertise is inherently transgressive[4]. As a result, science has become only one of many institutions that create and validate new knowledge, and scientists are finding it increasingly difficult to maintain and secure their position as credible and independent harbingers of truth in the marketplace of ideas[5].

The key to understanding the present transformation of science and the society in which it is embedded lies in recognising their co-evolution. It is not the impact of any specific parameter but their suggestive clustering and interdependent influence that lead to the complex interactions between science and other parts of society. Thus, for example, the emergence of new uncertainties is inherent in science and society alike, but they also feed on each other. There is a growing recognition of the potential of science and technology as the most powerful generator of the New, which is a tribute to the success of science, not evidence of its failure. Societies like our own have, at least in principle if not in practice, embraced innovation in a continuous drive to bring forth the New. They have acknowledged in a deep sense the necessity of living with uncertainties. But despite all this progress, the accumulation of uncertainties affecting social choices and behaviour, individual lifestyles, and personal and social identities, is unending. It reflects the equally inexorable increase in the number of notional options and actual choices, although these may be constrained in novel ways. None of these uncertainties can be limited

---

4 Nowotny, H., "Democratising expertise and socially robust knowledge", *Science and Public Policy*, 30, 2003, pp. 151-156.
5 Weingart, P., *Die Stunde der Wahrheit*, Velbrück, Weilerswist, 2001.

from the start or factored out. The generation of uncertainties is as inherent to, and endemic in, research as it is to contemporary life. However, the concrete arrangements to embrace them cheerfully or to make them tolerable, to thrive on them or constrain them, have to be worked out incessantly and concretely time and again. And it is here, in the realm of the uncertainties, that the natural sciences and the humanities again have to start a dialogue.

This is where we may return to the work of the two artists mentioned at the beginning and where – triggered by what art can achieve – the question arises of what can be done in practical terms to put the humanities and the social sciences on a converging path with the natural sciences, and *vice versa*. The main difference between what artists and an interdisciplinary team of scientists can achieve lies in the purpose of their work and their communication with the audience. Artists do not aim to provide definitive answers to questions posed by their imagination. They prefer to play with ambiguity, irony or with a host of other reactions and emotions that they try to elicit in the viewer. They seek to engage their audience in unforeseen and unforeseeable ways. Scientists seek to communicate as unambiguously as possible and look for answers that are as definite as they can temporarily be[6].

A converging research agenda must begin by sharing a minimum of mutual understanding and by developing a common language, however rudimentary. Three examples of this are already taking place on a small scale and in the specific local context of a research environment. They are experimental, and therefore precariously institutionalised; they might eventually cease to exist, but they might also be copied elsewhere, in a different form, to continue in

---

6 Nowotny, H., "Science in search of its audience", *Nova Acta Leopoldina*, NF 87, 325, 2003, pp. 211-215.
7 www.4sconference.org

unexpected ways.

The first example is taken from research carried out in the field of social studies of science. A research group in the life sciences invites an outsider to join. This is a person trained in social studies of science and technology, who may originally come from anthropology, sociology or even biology[7]. He or she participates in the research life of the group, and tries to understand what, as Clifford Geertz once said, "they (the natural scientists) are up to". On a daily level, the newcomer will try to be useful, perhaps learning technical skills. But he or she will also observe the ongoing social interactions, discuss the problems that arise and the meaning they attach to what they do. In short, the molecular biology or functional genomics lab becomes the new research territory for the social scientist, whereas the biologists will learn how their work is seen, interpreted and reflected back to them from a social science or humanities perspective.

The second example that is still in preparation takes us to a project of a group of scholars from the humanities. Typically, they are geographically dispersed and used to working in relative isolation. However, they have decided to form a network, with the Internet as their means of communication. They are interested in exploring the changes that occur when "writing" and "reading" no longer take place in the traditional way, but are transferred to the new medium of the Net. Among their questions is one that is deceptively simple: what is a library? To expand the scope of the investigation, the group decides to collaborate with life scientists who work with large data sets that have to be stored, processed and interpreted. The common grounds to be explored are therefore the similarities and differences in the concept of what constitutes a library, including in the life sciences. The importance of the library and information storage to both humanities and natural sciences is only now becoming better understood. The natural sciences increasingly have to cope with vast quantities of electronic information, its storage, distribution and access – problems the humanities are familiar with, although the technologies used by each of them may differ.

The third example, in which I am now involved, takes us back into the lab and the working environment of the life sciences[8]. A small group of motivated and gifted postdoctoral researchers has been awarded a fellowship for up to five years, which allows them to work wherever they choose. They continue their scientific work and career, but they integrate an additional dimension: society and how it relates to their science. They too are geographically dispersed, but they have opted for a loose cooperation, which allows them to meet and exchange views and results. They may also engage with people coming from social studies of science and technology and are encouraged to follow their curiosity in exploring the many known and unknown emerging interfaces between science and society.

For the research agendas to converge it is not sufficient to use slogans such as *Understanding the RNAissance,* nice as it sounds[9]; the changes have to be more profound. There is, however, an historical precedent. During the Renaissance, the intellectual élites of Europe realised that studying the works of ancient scholars from Greece and Rome and their successors in the Arab world was essential for understanding and building their present society[10]. Scholars and artists began to collaborate and exchange their interpretations of practical mathematics and ancient surveying methods, Latin orations and poetry. Artists ceased to see themselves as mere craftsmen and became aspiring scholars in their own right. The political and material environment was transformed to provide the political and economic space for the kind of patronage under which the new culture

9 www.nature.com/drugdisc
10 GRAFTON, A., « Remaking the Renaissance », *New York Review of Books,* 4 mars 1999, pp. 34-38.

flourished. The barriers between previously distinct realms such as visual arts and natural philosophy fell, and the creative potential of freely mixing scientific cultures with other forms and expressions of contemporary culture surged forward. It was a truly revolutionary time and the changes that took place during the Renaissance laid the foundations for Europe's lead in science and philosophy for many centuries to follow. Perhaps we should again ask a question similar to the one that Renaissance thinkers asked half a millennium ago: what has science to offer to life? It could lead to a renewal of the interactions between the humanities and the natural sciences if "life" is understood as a converging research agenda in which all the sciences participate.

LIFE SCIENCE, GOVERNANCE AND PUBLIC PARTICIPATION: THE NEW DILEMMAS OF DEMOCRACY

Massimiano Bucchi

Massimiano Bucchi teaches the sociology of science at the University of Trento (North Italy) and is a member for Italy of the Public Communication of Science and Technology International Scientific Committee. He has published five books among which are *Science and the media* (Routledge, 1998) and *Science in society* (Routledge, 2004, Mulino, 2002) and many essays in international reviews such as *Nature*, the *British Journal of the History of Science*, and *Public Understanding of Science*.

Recently, the debate on the topic of GMOs has become more intense, buoyed up by a series of events such as the heightening of tensions between the United States and Europe, the destruction of transgenic maize crops opted for by some local authorities in countries like Italy, and the approval of the Directives concerning the labelling of transgenic foods. The discussion has brought the issue of the relations between science, society and politics back into the news, with a series of interventions which offer a striking image of the current state of debate on these subjects.

Most of the interventions seem to be limited to considering the problem in terms of a clash between scientific expertise, on the one hand, and a scientifically "illiterate" public opinion, hostile to science as a matter of principle and bolstered by a "distorted" media portrayal and demagogic politics, on the other. They even go so far as to refer to Luddite lobbies opposing change – in most cases identified in a rather generic way – which would seem to "appeal to ancestral fears and unverifiable dreads, widely prevalent among public opinion[1]".

According to some influential opinion poll experts, the almost generalised opposition to GMOs by Italian citizens – recorded by the Eurobarometer and many other surveys – would be due to "media propagation, with the image that the mass media have rightly or wrongly given of GMOs. That is what has convinced most Italians to be unfavourable to them[2]".

But is that really how things are? Is it really just a problem of communication?

Communication, Risk and Interests

It is well known that the problem concerning the effects of communication has been keeping students of the mass media busy for over half a century; however, it is worth recalling that the direct influence of media information on attitudes – not to mention behaviour – has yet to be proved.

Several studies into the specific subject of biotechnologies have shown that disinformation cannot be considered as being solely responsible for the entire problem. In particular, more often than not, people who are not only most exposed to first-rate scientific content in the media, but also are the most informed about biotechnologies, are no less critical for all that – in fact, in some cases, the opposite is true: the most informed are also the most sceptical[3]. It is therefore not enough to give more widespread and better communication to the risk to make it

---

1 CARRUBBA, S., "Contro le lobby anti-innovazione" (Against the lobby of opponents of change), *Sole 24 Ore*, 18 May 2003.
2 MANNHEIMER, R., "Via libera ai cibi transgenici? Due italiani su tre dicono no" (No obstacle in the path of transgenics food? Two Italians in three say no), *Il Corriere della Sera*, 4 August 2003.
3 See, for example, GASKELL, G. & BAUER, M. (Eds), *Biotechnology 1996-2000: the Years of Controversy*, London, The Science Museum, 2001; BUCCHI, M. & NERESINI, F., "Biotech remains unloved by the more informed", *Nature*, vol. 416. 21 March 2002, p. 261.

acceptable; otherwise the warning "Smoking seriously damages your health", which is marked on all cigarette packets would be enough to discourage millions of smokers. Furthermore, if more information and greater interest by public opinion are always desirable, on how many issues on today's political agenda – from the Middle East to the economic recession – do we really have accurate and full information? A more sophisticated variant of this subject places the problem in the relationship between science and politics – "a dialogue of the deaf", as the science philosopher Riccardo Viale defines it[4]. The mistake of politics, for Viale, is to "refuse to consider institutional science, namely science as expressed in the main international science reviews, as the only source of knowledge about phenomena of the physical and biological world. It therefore implicitly accepts that the choice of sources of knowledge and the ways in which that knowledge is produced are guided by social and cultural reasons." In short, the political class, wishing to please public opinion for electoral purposes, does not give science a suitable hearing.

This is an interesting argument because it actually reproaches politics for not operating like science but following its own institutional rules, for being more attentive to social and cultural reasons than to those of the *peer reviews*. The argument is very similar to the one expressed by the former Italian Minister of Health, Umberto Veronesi, who proposed that an "upper house" should be set up, composed of independent intellectuals rather than politicians, to cope with difficult issues such as those arising from biotechnologies[5].

Lastly, there is the argument about interests. The opposition to biotechnologies – and in particular to GMOs – would seem to stem

---

4 VIALE, R., "Scienza e politica dialogo tra sordi" (Science and politics, dialogue of the deaf), *La Stampa*, 6 August 2003.
5 VERONESI, U., "Una camera alta per etica e scienza" (An upper house for ethics and science), *Il Corriere della Sera*, 19 May 2003. A detailed response in BASSETTI, P., "Conciliare etica e scienza" (Reconciling ethics and science), *Il Corriere della Sera*, 28 May 2003.

from lobbies opposing change who have an interest in promoting organic farming at the expense of transgenics. Now, there is no doubt that economic interests can be linked to any position on an issue. It is not clear, however, where this will lead us, unless the intention is to demonise the role played by interests in scientific research and technological innovation. It is obvious that some people get rich by producing medicines, but that does not make such medicines less useful to the patients. If it had not been for the interests of French breeders, perhaps Pasteur would not have been able to carry out some of his most important experiments on vaccines. We would not have the aspirin or the bouillon cube either. It is well known, moreover, that interests represent a significant element in that process which "selects" the ones which will survive among the many technically plausible innovations[6].

How Relations between Science, Society and Politics are changing

Similar interpretations risk underestimating the thorough change that the relations between science, society and politics are undergoing. This change sets a dilemma which, on the contrary, is more far-reaching, more serious and of increasing topical interest. In fact, public debate about agrifood biotechnologies only represents one example of it, even though it is a striking one: the question is how to reconcile forms of democracy with the *governance* of problems of high technical and scientific complexity? It is a dilemma which emerged in a particularly forceful way during the second half of the nineties, in the wake of events such as the emergencies related to BSE, the first

---

6 As an example, see the recent review by NOSENGO, *L'estinzione dei Tecnosauri. Storie di tecnologie che non ce l'hanno fatta* (The extinction of the Technosaurs. Tales of technologies which didn't make it), Milano, Sironi, 2003.

cloning experiments using adult animal cells and, obviously, biotechnologies themselves in a broad sense. Various institutional bodies, starting with the European Commission, currently consider the solution to this dilemma (and, more generally, a greater dialogue between scientific researchers and public opinion) as one of the priority objectives of policies in the area of science and technology[7].

Tackling – and expecting to solve – this dilemma by resorting to a traditional conception of the relations between science, society and politics means not really grasping just how these dimensions and their relations have changed. However, they have not only changed insofar as there have been an increased number of political problems of a highly complex technical and scientific nature. That is only the surface of the problem. Even in the past, politics had to tackle this kind of issues. Give a thought to the situation of the United States government when it had to decide whether or not to build and use the first atom bomb. The difference was that the relationship between politics and scientific expertise – as was clearly described by Snow – could and should take place "behind closed doors", i.e. away from public debate[8]. That is why there was no public discussion for years – as is the current case, on the contrary, concerning GMOs – as to whether it was more or less dangerous to use herbicides or fungicides in farming, so that it was the book by a writer, Rachel Carson, which made the issue break the news in the United States[9]. Yet the matter possibly involved an innovation that was at least potentially just as controversial?

Today, however, such a "closed" procedure – which partly meets the wishes of Viale and Veronesi, among other people, namely politics, which turns to the man of science as an advisor – is no longer a course of action for various reasons. The first one is the role assumed

---

7 See the European Commission's plan of action, "Science and Society":
http ://europa.eu.int/comm/research/science-society/pdf/ss_ap_en.pdf
8 Snow. C.P., *Science and Government*, Cambridge, 1960.
9 Carson, R., *Printemps silencieux*, Paris, Plon, 1963.

by the media which makes it impossible to handle such a relationship within the closed halls of politics. The second reason is the actual transformation of scientific expertise, especially of its perception by public opinion and by politics itself. The current question is not so much wanting or not wanting to rely on scientific experts, the question is rather: "which experts?" Politics, even more so than public opinion, is currently in a situation where it is bewildered by a diversity of experts who frequently offer very different assessments one from another. Environmentalist organisations actually have their own trustworthy scientific experts who speak on the organisations' behalf whenever matters such as the greenhouse effect or GMOs are being discussed. What should "responsible political leaders" – as influential commentators often refer to them – do when confronted with the decision whether or not to destroy crops of transgenic maize? Should they carry out a majority survey among the researchers concerned with that subject? Should they consult the latest issues of the magazine *Nature*, where studies have been published which rule out the dangerousness of GMOs, along with others which do not exclude such a hypothesis? Should they believe the *Istituto Superiore di Sanità* (Italian National Institute of Health) or the *Consiglio Superiore di Sanità* (Italian High Health Council) which, in 1999, at the request of the Ministry of Health, issued conflicting information about some transgenic products in a short space of time? As an example, and just to make it clear that the problem does not only concern biotechnologies, the commission set up by the British Government to ascertain whether the current "internal" radiation risks due to the inhalation or ingestion of radioactive substances released into the atmosphere should be revised, is still paralysed by the different hypotheses offered by the various experts consulted[10].

---

10 HOGAN, J., "Experts can't agree on internal radiation risk", *New Scientist*, 19 July 2003.

Whether or not there is still a "scientific community" today, in the traditional sense – in other words, a set of actors marked by great internal homogeneity and sharing a good deal of specific aims and values – is too vast a question to be dealt with in these pages[11]. What is important, however, is that the public and the political world currently perceive it as being increasingly divided and internally not homogeneous, while finding it hard to identify incontrovertible spokesmen. Incidentally, the "closed" relationship between science and politics, described by Snow as typical of the *big science* which sprang up in the inter-war years, was not free from risks, the most important being the fact that the experts to whom people turned, by maintaining an absolute reserve, were actually removed, to a large extent, from any control by the reference scientific community.

● ● ● Trust in Science, Ethics and Responsibility

It would be easy to mistake this perception for a loss of trust in science and researchers, by falling back on the *cliché* that holds that Italian public opinion is insensitive, hostile and suspicious of research[12]. However, most recent studies, conducted both at national level and compared with essential findings at European level, concerning the attitudes of public opinion towards science and biotechnologies[13] in particular, tell us that scientists are back at the top of the list of trustworthy categories and institutionson biotechnologies. According to data from the most recent survey by the *Observa Research Centre*, scientific institutions are the source to which Italians give the greatest

---

11 Or this theme see, for exemple, ZIMAN, J., *Real Science. What it is, and what it means*, Cambridge, Cambridge University Press, 2000; NOWOTNY, H., et al., *Re-Thinking Science. Knowledge and the Public in an Age of Uncertainty*, Cambridge, Polity Press, 2001.
12 For a critique of this point of view, see NERESINI, F., "Siamo una democrazia poco scienti-fica?", *Palomar*, XIV, 2003, pp. 82-93.
13 *Europeans and Biotechnology in 2002*. Eurobarometer 58.2, 2003.

credit on the subject of biotechnologies[14]. In any case, the problem is otherwise: it is worth pointing out that over two-thirds (68.6%) of the same people interviewed consider the scientific community as "divided" on the subject of GMOs. Lastly, an almost identical portion describes science as being "interested".

In other words, what has been lost in recent years is not so much trust in science or in scientists, but an image of "neutral" and disinterested science which had been cultivated for a long time in the public sphere. This change questions the validity of a further aspect of technocratic proposals, such as the proposal for an "upper house" for science composed of "independent" experts[15]. Independent of what and whom? Of political power? Of environmentalist organisations and consumer associations? Of businesses? The proposal underestimates the fact that we are no longer dealing with post-war *big science*, mainly played out between academy and political decision-making, but with a science characterised by a new "academic-industrial-government complex" in which announcements of scientific discoveries send private company quotations shooting sky high. Nowadays, specialised scientific reviews themselves are increasingly often in an awkward situation when it comes to publishing findings of research funded by companies in the pharmaceutical or agrifood industry[16].

These aspects are interwoven with the issue of ethics. There are widespread proposals for the setting up of ad hoc ethical committees as a solution to the above-mentioned dilemmas and that is partly the

---

14 *Biotecnologie: Democrazia e governo dell'innovazione. Terzo rapporto su Biotecnologie e opinione pubblica in Italia* (Biotechnologies: Democracy and governance of change. Third report on Biotechnologies and public opinion in Italy), www.observanet.it . The study was partly supported by Fondazione Giannino Bassetti for Responsibility in Innovation.

15 VERONESI, U., *op. cit.*

16 Cf. for example, VAN KOLFSCHOOTEN, F., "Can you believe what you read?", *Nature*, 416, 2002, pp. 360-3. More recently, a group of food scientists signed a protest against the failure by the Nature Biotechnology magazine to point out a conflict of interests between the authors of an article who asserted that GMO foods were not dangerous; the letter points out how 11 of the 18 authors had received funds from a company operating in the agrifood biotechnology industry.

message of the proposal by Veronesi who, moreover, advocates an ethics specific to scientists capable of safeguarding society. This solution, however, also seems unsuited to a framework which has undergone such a thorough change. Hans Jonas asserted, for example, how the traditional ethical reflection – which used to be related to short-range actions in time and in space and considered human action on non-human objects as being unimportant – is of little use today, at a time when science and technology are making available to us the possibility of actions with previously unthinkable consequences and importance[17]. That is the central focus which the German scholar assigns to the concept of responsibility. But the traditional concept of responsibility is in crisis in the face of contemporary science. The articles reporting the results of the mapping of the human genome in the *Nature* and *Science* magazines in 2001 were signed respectively by 275 and 250 authors. Who among them is capable of assuming responsibility for the impact of that research? Perhaps the shareholders of the private company Celera, which shared pre-eminence for it with the public consortium? Or the politicians who decided, in 1993, to provide massive funding for it by giving priority to it over other lines of research?

And who assumes responsibility for deciding whether on not transgenic maize should be destroyed to avoid contaminating the non-transgenic variety? Unlike some of the commentators mentioned, I believe that politicians would look forward to unloading this responsibility on to the expert on duty. The problem is that it would not work. It would not work because of what has been previously said concerning the changes undergone by scientific expertise and the way in which it is perceived by society. It would not work because science has lost its holy aura as an area of neutral and disinterested action. Lastly, it would fail because the non-specialist public – another epoch-

---

17 JONAS, H., *Das Prinzip Verantwortung*, Frankfurt, Insel, 1979.

making change – is increasingly clamouring to participate, to be involved, to have a say in the matter even where highly complex, technical and scientific issues are concerned. We have had several examples of this since the nineties – from the mobilisations of associations of patients and homosexuals, which deeply influenced the pathway followed by research in the field of AIDS, especially in the United States, to the Di Bella case which made us understand, in Italy also, just how permeable the boundaries between science, society and politics had become[18]. We can turn up our noses to this phenomenon, but the same conditions which today make public discussion about biotechnologies possible have given the means to support and finance – on the basis of such "social and cultural reasons" which many people deplore – research in cancer and AIDS in ways and on a scale never previously seen in the history of medical research.

Accusing the media of sensationalism, public opinion of being obtuse, politicians of demagogy and insensitivity to the reasons of science, is taking a short cut through the illusion of still talking rationally about science (and the social role of science) as it was in the middle of the last century or even in the middle of the 18th century. Being outraged because reasons of a cultural nature, "national interests and Community policies" hang over the decision concerning transgenic maize[19] means forgetting that GMOs, BSE and SARS are no longer scientific issues, but scientific-political-social issues and science has entered politics and business in no less a way than politics and business have entered science. Even in the legal field, a technicalised view of the application of scientific knowledge and more generally of the relations between science and law, has been superseded over the

---

18 In 1997, in Italy, pressure from patients associations forced the specialists – who were very sceptical – to try alternative therapies for cancer treatment.
19 Meldolesi, A., "La scienza partecipativa ? Un bluff" (Participatory science? A bluff), *Il Riformista*, 20 june 2003.

last decade by an outlook which no longer only sees law as using the instruments of scientific research, but as taking a significant part in it, for example by laying down how scientific discovery may be patented, who may be considered as a scientific expert, and even what counts as "mathematical proof"[20].

●●● Conclusion

Many signs lead us to assume that we are faced with a crisis of legitimation of the procedures linking scientific expertise, decision-making and political representation rather than hostility towards biotechnology or a crisis of the legitimation of science as such.

On a subject such as biotechnologies, neither the beaten tracks of politics nor technocratic solutions appear viable anymore . The role assumed by the media, political and social changes and, above all, the growing demand for involvement which the public expresses towards issues with a high "expert" content, require new strategies and new forms of participation. It is certainly difficult to say what kind of strategies and what kind of forms of participation. Issues such as AIDS or BSE have contributed to highlighting the relevance – both for scholars and for those involved at the operational level – of various forms of citizen involvement in decisions related to science and technology, such as "consensus conferences", "voting conferences", and "scenario workshops". These formats have been experimented with since the eighties in countries like Denmark before being introduced in many other contexts. The model of the "consensus conference", for example, provides for a dialogue between citizens and experts, usually open to the public, to media spokesmen and to

---

20 See for example: MacKENZIE, D., "Negotiating Arithmetic, Constructing Proof: The Sociology of Mathematicsand Information Technology", *Social Studies of Science*, XXIII, pp. 37-65 ; TALLACCHINI, M.C., "Democratizzazione della scienza e brevetti tecnologici", in BERNASCONI, C. & al. (Eds), *Cellule e Genomi*, Pavia, 2003, pp. 63-84.

the political world, on a controversial scientific or technological topic. A panel composed of a number of citizens, which generally varies between ten and 30 ("citizen panel"), selected in a random way, proposes the key questions and takes part in the selection of the panel of experts ("expert panel"). The latter panel is composed of experts with communication skills, a thorough knowledge of the sector under consideration, and representative of different positions. Once the experts have been heard, under the supervision of an "advisory committee" whose task is to ensure that the process follows the rules of democracy and transparency, the citizen panel must draw up a document with its own conclusions and recommendations[21].

The development of these panels has been particularly lively in the operational area of the assessment of the potential impact of technological innovations – the so-called *Technology Assessment* – setting up a specific one currently referred to as *Participatory Technology Assessment*. Forms of consultation and involvement of citizens in decisions about technological innovations now represent a stable component in the activity of public institutions, such as the Danish *Board of Technology* or the Swiss Council for Science and Technology[22].

Analyses of the effectiveness of such models do not provide unambiguous conclusions, even if it is clear that it is not enough just to bring together citizens, experts and policy-makers around a table to unblock debates as complex, for example, as the one about biotechnologies. In comparison with an instrument such as the referendum, the approach being experimented with until now offers the advantage of being able to involve citizens from the initial policy process phases by encouraging contact with experts before positions

---

21 Joss, S. (Ed.), "Public Participation in Science and Technology", numéro spécial de *Science and Public Policy*, XXVI, 1999, pp. 290-293.
22 Joss, S. et Bellucci, S. (Éds), *Participatory Technology Assessment*, Westminster, 2002.

polarise; moreover, these methods necessarily involve a limited number of citizens and are not usually binding on political decision-making. There are also a series of variables to be taken into account, from the scale of the problems discussed – participatory models have given the most concrete results on a local basis up until now – to the institutional configuration of the participatory event: its management by a government agency (as is often inevitable for problems of resources and organisation) may re-propose the limits of an approach in which citizens are relegated to a mainly passive role[23].

Of course, it is naïve to think that these forms of involvement in themselves can represent an immediate solution to the new dilemmas of political decision-making on the subject of innovation. Perhaps it is just as naïve, however, to think that the problem can be simply crushed under "a mountain of scientific data"[24].

And not only for the reasons which have been mentioned but because this reduction of the sorting policy of scientific expertise acts primarily to the detriment of science itself! The increasingly heavy charge of responsibility in the face of public opinion transforms science into an easy scapegoat as soon as an emergency threatens which escapes its control.

This "humiliation of politics", according to Bruno Latour, is the result of a philosophical tradition summarised in the allegory of the *Cavern* by Plato: to reach the truth, the scientist has to free himself from the "tyranny of social conventions, of public life, of politics, of subjective feelings, of popular agitation[25]". Such a tradition, without which science would probably have not been in a position to develop and institutionalise , has currently reached a turning point. In principle, the moralism about "contaminations" with social interests and

---

23 IRWIN, A., "Constructing the Scientific Citizen: Science and Democracy in the Biosciences", *Public Understanding of Science*, X, 2001, pp. 1-18.
24 MENDOLESI, A., *op. cit.*
25 LATOUR, B., *Politiques de la nature*, Paris, Editions La Découverte, 1999.

pressures risks really endangering the fundamental role of science and technology in contemporary society. Whether it pleases us or not, science today has to take into account – many researchers and many institutions are already doing so –   conflicts of interest, pressure groups and associations, and the media.

We are actually faced with new decision-making issues and challenges which cannot be dealt with solely within politics or solely within science. It involves understanding how to reconcile the challenges of expert knowledge and innovation with democratic procedures and with their actual renewal in a society which is becoming increasingly drenched with social-political-scientific-technological objects and decisions and in which each individual element is less and less separable from each other.

Science Fiction: the cultural Spin-offs from the Life Sciences

Since the time of Mary Shelley, the author of *Frankenstein*, fictional writers have had a strong influence on popular images of biological science. Her literary descendants have been many, including H.G. Wells, Jack London, Karel Capek and Aldous Huxley. Biology has subsequently served as an inspiration for filmmakers, pulp writers, essayists, journalists and other commentators. How does the production of science fiction as a cultural phenomenon relate to the growth of biological knowledge and its subject matter: the origins, nature, and continual evolution of life? Across cultures, people have made use of the same ideas to tap into biology and imagine countless possible and impossible scenarios for the human condition and the destiny of humanity.

Who are the main consumers of science fiction: non-scientists or scientists? Does science fiction help to popularise science, or does it just spread misconceptions? And what effect does reading science fiction have on scientists? Does it encourage new ideas and stimulate creative thinking? If science were a religion, would science fiction be its mythology? Many science fiction writers have proved extremely successful in reaching a broad public, while scientists often admit to profound difficulties in communicating science. Have science fiction writers, with their proven ability to penetrate and stimulate public imagination, got anything to teach scientists? Suspended, as they are, between nature and culture, the life sciences provide fiction writers, dramatists and filmmakers with a rich source of material for weaving their popular, science-inspired visionary tales. The final session of the conference will examine the many and complex relationships between science and fiction.

THE DAY AFTER TOMORROW

Christopher Bigsby

Christopher Bigsby is professor of American studies at the University of East Anglia (UK). He has published over thirty books about the various aspects of the English and American culture such as: Afro-American literature, theatre drama and popular culture. His last novel, *Beautiful dreamer*, narrates the story of two men – the first black, the second white – which an act of terrible violence will force them to run away.

Time present and time past
Are both perhaps present in time future,
And the future contained in time past.

T.S. ELIOT, *Burnt Norton*

Dr Johnson once observed that "whatever withdraws us from the power of our senses, whatever makes the past, the distant, or the future, predominate over the present, advances us in the dignity of thinking beings". When architects devise building complexes they often plot the course of pathways by calculating likely use, points of departure and arrival. When the buildings are up and the paths laid, people often choose quite different routes, marking them out in mud as they steadfastly refuse the rational way to go from A to B. Architects have a word for these. They call them lines of desire. We advance along the lines of our desire, often refusing the rational, inventing the future we wish to inhabit, and in many ways it is the future tense which defines us as human animals. As George Steiner has reminded us, the concept of tomorrow is not one we share with other creatures. Only human beings, he suggests, can contemplate the day after their death, hence the vindictive pleasure with which some draw up their wills. This is even truer of the future perfect, that tense which asserts that something will have occurred. In Ursula LeGuin's *Always Coming Home* we learn of a people who *"might be going to have lived in California"*. There is, as William Blake asserted, something distinctive in those *"who present, past, and future sees"*. It is certainly something that scientists and science fiction writers share. If we do not lean into the future we are back in Eden, an eternal present with the fruit still on the tree.

● ● ● Science and Science Fiction

Once, science fiction was thought to be the illegitimate offspring of science, a bastard child sent swiftly to some distant relative in the country. Scientists were suspicious and so were writers of what bookstores today like to call "literature". Yet in truth "real" writers were always writers of science fiction and fantasy. Gulliver travelled to other lands, though not so exotic that they lacked scientists. Anatole France stared into the future. Edgar Allan Poe and Jack London more than flirted with the form. Ironically, though, the time came when science fiction writers themselves became suspicious of those who entered their territory and welcomed them, such as North Korea might a tourist inadvertently wandering across its frontier.

Today the distinction no longer holds. Major authors move in and out of science fiction as though it were indeed not different in kind: J.G. Ballard, Kurt Vonnegut, Ursula Le Guin, Margaret Atwood, Doris Lessing. As Ursula Le Guin once said, "the funny thing.. the heartbreaking thing... is that within the little science-fiction community, the self-ghettoised science-fiction community that doesn't want anything to do with the outside world, they don't even talk about Lessing and Atwood. And these days they are rather uncomfortable with me because I have "betrayed" them. I have refused to stay only in the ghetto. I want to have the freedom of the city. So it is a bit of a mess on both sides. There is a lot of territoriality; people spraying and marking their boundaries."

Scientists and science-fiction writers have done a fair amount of spraying of their own and their two countries do have different borders. But those borders are far from secure. There are a number of defections and certainly a fair amount of cross-border trade.

Several decades ago, C.P. Snow famously referred to the two cultures, with science on one side and the humanities on the other, staring at one another in bewilderment. In 2003, the President of the British Academy saw the same division. And yet is it so clear? Edward Teller was a near concert-level pianist. Nobel Prize winner Harry Kroto, at least when I knew him as a fellow student, was an accomplished graphic artist. Robert Winston's book on the human mind contains chapters on *Hamlet* and Pirandello. Steve Jones' book on the Origin of Species updated, is entitled *Almost Like a Whale*[1], a deliberate play on a line from *Hamlet* and begins with what, with some justification, he calls two of the worst lines of English poetry, written in 1799 by John Hookham Frere:

> *The feather'd race with pinions skim the air —*
> *Not so the mackerel, and still less the bear!*

Skim the index of Jones' book and you will find references to Goethe, Jack London, Herman Melville and George Bernard Shaw. And he reminds us that Darwin himself recorded his own reading of *Hamlet*, *Othello*, Jane Austen, and *Robinson Crusoe*.

---

1 JONES, S., *Almost Like a Whale*, New York, 1999.

Of course, I am teetering on the edge of a terrible solecism. It is only surprising that scientists are lovers of literature, music and art if for some unexamined and unsubstantiated reason one assumes they are incapable of such. We are various. The British education system may require 15-year-olds to decide whether they will study the arts or science but this kind of Manicheanism has little justification and thankfully does not extend to many other countries.

The same family produced the scientist T.H. Huxley, the novelist and science-fiction writer Aldous Huxley, and the writer/biologist Sir Julian Huxley who while being Director-General of UNESCO, also turned his hand to science fiction. Not for them some absolute divide. Incidentally or, rather, not so incidentally, one of T.H. Huxley's pupils at the Normal School of Science in South Kensington was a young man called H.G. Wells who studied biology with Huxley, learning enough to write his Darwinian novel, *The Time Machine* (1895). Meanwhile, his reading of Frederick Soddy's *The Interpretation of Radium* led to his vision of a future nuclear war, *The World Set on Fire*. That book was published in 1914, a good year to be contemplating Armageddon.

Writers are drawn to science like wasps to ripe plums, just as they are drawn to history, biography, and autobiography. They are magpies attracted by glitter, and science glitters. John Fowles, who staged a debate about Darwin in his novel *The French Lieutenant's Woman*, is also a naturalist, an amateur palaeontologist and called a recent collection of essays *Wormholes*. This last was not, I think, because he is an enthusiast of *Star Trek* but because he saw in the idea of hypothetical interconnections between widely separated regions of space-time a useful metaphor, as the astrophysicists who devised the term had themselves turned to metaphor in order to explain the concept to themselves and others. And how often does the scientist, seeking to explain an idea, begin his or her account with an image or with the word, "imagine".

We are used to writers staging raids on science but scientists, as suggested above, are also drawn to written science fiction, even writing it. For some, no doubt, this is a little like playing tennis with a Stradivarius but among those have chosen to do so are Norbert Wierer, Otto Frisch and Leo Szilard. Robert L. Forward, a research scientist and science fiction writer, has explained that "I don't ... write science fiction. I just write a scientific paper about some strange place... and by the time I have the science correct... the science has written the fiction."

Imagine a poker game between Sir Isaac Newton, Albert Einstein and Stephen Hawking. Well, as it happens there is no need to imagine it. It actually took place, if you allow a little licence for the word "actually". Isaac Newton and Albert Einstein were holograms or, rather, actors playing holograms. Stephen Hawking was real enough, if someone could be said to be real while acting, even if he was acting himself. The fact is that this scene took place on the "holodeck" of the *USS Enterprise* in an episode of *Star Trek: The Next Generation*.

Hawking (who, being a man of good taste, was also to appear in *The Simpsons*) had, it turned out, long been a fan. Now he stepped into this factitious future as, in his mind, he had stepped into another future, equally factitious but to be tested for its theoretical possibilities.

Before the scene was shot he was taken on a tour of the "engine room" and shown the warp engine that could drive the ship faster than light. *"I'm working on that right now"*, he explained as, presumably, Einstein – hologram, or not – revolved in his grave, albeit at something less than the speed of light. On the other hand, if *Star Trek* had, to say the least, its improbabilities, it also sought inspiration in the work of scientists. Science fiction and science are, if not brothers, then in some respects kissing cousins.

Douglas Adams was a writer of somewhat surreal science fiction. He cut his teeth on *Dr Who* which, for those who have never heard of it, was a British television science fiction programme whose effects were special only in the sense that they never got much beyond the cardboard and Scotch tape variety. He was fascinated by science and technology and was not averse to checking his science with card-carrying scientists. At the same time, his science and technology were never quite subject to the normal laws of physics or any other science, except perhaps for displaying a kind of aesthetic entropy. Stephen Hawking wished to be in a science fiction film; Adams wanted to become a scientist, even thinking of returning to university to read zoology. They were plainly different from one another but equally plainly had discovered they could breathe the same air. It is somehow appropriate that the zoologist Richard Dawkins is married to an actress who once appeared in a *Dr Who* episode written by Douglas Adams and that the two men were friends and admirers until Adams' premature death.

Hugo Gernsback, whose *Amazing Stories* and *Astounding Science Fiction* had been breeding grounds for science fiction writers, had himself had a technological education and began with a science magazine. For some years science and science fiction co-existed in the same magazines. Soon, though, the fiction began to predominate. In a society obsessed with presenting itself as the harbinger of the future, for which the word "progress" was a slogan (the 1933 Chicago Exhibition celebrated "A Century of Progress"), tomorrow was the lure and science seemed a little slow in getting us there. This, after all, was, as Henry Luce asserted, to be "The American Century". Walt Disney, despairing of science's tardiness, chose to have his head cryogenically stored and awaits in the deep winter of cryogenic suspension for the day the scientists catch up with the science fiction writers.

We forget how recent is the assumption that the disciplines are closed to one another. In the 19th century, public debates over science were assumed to be available to most intelligent readers. The obligation to explain has always been felt by the scientist, and the hunger to understand has always been felt by those aware of the relevance of that science to themselves. And although the inner processes and procedures, the theoretical underpinnings, of science may tax the average reader, I doubt there has ever been a time when science has been such a primary concern. The evidence for this lies in the immense sales of books in which scientists endeavour to open the doors into their world because it is also our world. The classic example is *A Brief History of Time*, but there are if not countless others then a large number while science fiction commands a separate column in the *New York Times Book Review*. The title of Richard Dawkins's Chair at Oxford is the Charles Sominyi Professor of the Public Understanding of Science, and Richard Dawkins is both a Fellow of the Royal Society and a Fellow of the Royal Society of Literature. Meanwhile, a conversation between two nuclear physicists became the basis of an international success in the theatre in the form of Michael Frayn's play, *Copenhagen*, while Tom Stoppard turned to the Second Law of Thermodynamics in *Arcadia*.

There has always been a counterforce, however, a suspicion of change, of futures plotted either by scientists or writers. Henry Wadsworth Longfellow insisted on the central necessity to seize the moment, live in the present:

> *Trust no Future, howe'er pleasant!*
> *Let the dead Past bury its dead!*
> *Act – act in the living Present!*
> *Heart within, and God o'erhead.*

Carpe diem. But what is this present, this past and this future? As Eliot hints, the past inheres in the present as the present is compacted with the future like a child in the womb, heart beating against tomorrow's dawn. And is time so easily delineated?

Today, there are those who from our perspective live in the past.

Thomas Hardy, writing in *Far From the Madding Crowd*, observed that "Five decades hardly modified the cut of a gaiter, the embroidery of a smock-frock by the breadth of a hair. Ten generations failed to alter the turn of a single phrase. In these Wessex nooks the busy outsider's ancient times are only old; his old times are still new; his present is futurity." As we reach for the stars, others still draw their water from polluted wells and regard literacy as a form of necromancy. Mass murder is committed in the name of atavistic creeds.

Ursula Le Guin has also spoken up for the present, but from a rather different perspective than Longfellow, remarking that "the fact is that I don't believe in the future. It doesn't exist. All we have got is now and it is what we make of now that matters. That is where the imagination comes in. With the imagination you can make now into anything you please and that may lead you to a different future than the one you would have had if you simply projected the present forward. One thing fiction does is offer endless options." It may be that scientists would make a similar claim, for science develops not only by logical extensions of current knowledge but also by leaps of the imagination with the power to transform the future. It explores different options if not with quite the promiscuity of the writer.

I see, you see

Moby Dick famously does not begin with the words, "Call me Ishmael." It opens with a passage ostensibly by a sub-sub-librarian which offers what amounts to a scientist's approach to cytology. We learn the size, weight and reproductive habits of whales. We are offered a taxonomy, the fruit of observations, dissections, analyses. The rest of the book is concerned to show how distant this is from an understanding of how we constitute the world we observe.

As Pip the cabin boy observes of those who come to look at the design on a gold coin nailed to the mast: "*I see, you see, he sees.*" Moby Dick bears the impress of those who observe it. Ahab looks beyond the pasteboard mask. He is concerned with greater profundities than Starbuck's domestic sensibility can envisage or the scientist's detached observation suggest. Emile Zola bid the writer to be as cold as a vivisectionist at a lecture in order to aspire to the objectivity of science quite as if novel writing were a second-order activity. Melville tells a different story. He is making a case for literature, for what can never be harpooned, pinned down. Is this mystification or an opting for mystery, opacity, ambiguity, the Protean, the generative strength and power of the unknowable? As Harold Pinter remarked when asked for details of his characters, the desire for verification is understandable but it is a desire he could not satisfy any more than we can in daily life when we encounter strangers, or even friends and lovers, whose inner lives are closed to us. Perhaps the slogan should be, "No taxonomy without misrepresentation".

On the other hand, Richard Dawkins has insisted that to know the genome of an animal is not the same as to understand that animal, and to some degree that was what Melville was proposing. But Dawkins has no patience with the idea of mystery which he sees as elevating ignorance to the position of higher truth. Awe, he understands; mystery is a form of infection. And that is the root of his suspicion, deepening to fierce antagonism, towards religion, of which more in a moment.

Steve Jones has observed that both Darwin and Melville say a lot about whales, "but to rely on The Origin in a philosophy class is like using Moby Dick as a zoology text. One is fact, the other metaphor. The Arts Faculty often find it hard to tell the difference." In truth, I think Arts Faculties are alive to this. More importantly, so was Melville. He was in fact precisely making a case for metaphor while seeing the danger of metaphor in that Ahab goes to his death precisely because he transforms fact into image. Steve Jones would probably think this poetic justice. It may be true, as Jones also says of the social

195

sciences' use of evolution that it is a convenient platform upon which to deposit badly digested ideas. Many a writer does indeed sit down at the table of science and, confronted with baked Alaska, tastes the warm meringue unaware of the cold secret at its centre. Why let fact spoil a good story. Dawkins and others have done a justifiable demolition job on the ignorant appropriation of science by literary critics, sociologists, and psychologists, several of whom are fluent in gibberish in more than one language.

Yet there is perhaps a pedantry in asking that the writer should be over-respectful of scientific plausibility, something demanded by writers of what is known as "hard" science fiction. In a recent book (not by a science fiction writer), the fact that stars appear to move steadily past the windows of the Starship *Enterprise* was advanced as evidence of the dumbing down of America since stars do not actually do this. Aside from the fact that the *Enterprise* does not actually exist, and that, although "Klingon" is now spoken by more people than speak Esperanto, neither do Klingons, seems not to have worried the author. Nor, one presumes, does the fact that in Shakespeare we can move through time and space with nothing but language to carry us there. No respect for the laws of physics in *Anthony and Cleopatra*.

Truth in literature is of a different kind. There is a licence to speculate, invent, entertain. That is what we pay our money for. Men and women did not gather in "great halls" more than a millennium ago to be told that green dragons did not exist. And, like dreams, such stories are, indeed, sometimes ways in which we process concerns. The Cold War bred science fiction films about the invasion of America by aliens. Ray Bradbury's *Fahrenheit 451* registered a threat of book burning in 1953 at the height of the McCarthy period as *The Thing* acknowledged the fear of a communist takeover. In the 1950s, *The Day the Earth Stood Still* addressed alarm at the idea of nuclear war. *The Matrix*, the *Terminator* trilogy, both set in, or invoking, the future, respond to current alarms about our increasing reliance on technology. *Minority Report* offers a response to fears about intrusive security agencies with access to new technologies, while the *Alien* series responded to the threat of the military-industrial complex. An amoral business wedded to a science which lends itself to exploitation fired the imagination of Michael Crichton who, in *Jurassic Park*, took us back to the future. And science fiction, as in *Blade Runner* or, indeed, *Back to the Future: Part III* (set in the 19th century), often contains regressive features, creates retro environments seemingly located in the future but which are in fact in the present and past.

There is, perhaps, a case for regarding science fiction as a branch of the historical novel. Nothing, after all, ages so quickly as the future. Look at the original Buck Rogers films and you will see not the future but the past. In the original *Star Trek* you can smell the 1970s and 80s. In some ways, science fiction is the most conservative of forms, its roots in fantasy being clear enough. However innovative the technology, the social organisation is anything but radical. The future, it seems, is as peopled with kings, princesses and dukes as fairy tales and fantasy fiction. Set in the future, science fiction is written in the past tense and engages the present. It is what will have happened. Edward Bellamy's *Looking Backwards*, written in the 19th century, set in the 20th, commented on the 19th.

● ● ● Saint Dawkins

We are asked, "if science were a religion would science fiction be its mythology?". It is a dangerous question to ask, certainly in the presence of scientists. For Richard Dawkins, not a man to sit on the fence, religion is a "mind parasite[2]", although he has noted, with a mixture of amusement and horror, that there is supposedly an internet religion called the "Church of Virus" whose saints include both

---

2 DAWKINS, R., *A Devil's Chaplain: Selected Essays*, London, 2003.

Darwin and Dawkins himself. Religion, to Dawkins, is an unthinking replication of baseless assumptions. Faith is specifically counterposed to evidence.

Science may have some of the trappings of religion – its own rituals, interpretations, sacred texts and priesthood – but it lacks precisely that commitment to the irrational which religion claims as its foundation, although scientists are as liable as anyone to cling to interpretations long after the evidence ceases to support it. Scientist, priest, we are all human. Doubt, although fundamental to science, is seen as a threat to the authority of religion. To insist that there were no such things as witches was to invite being hanged on Boston Common or anywhere in the then Christian world. To obey injunctions with no discernible basis in reason was and is seen as evidence of true faith setting one on the path to sainthood. Few if any scientists have killed or tortured one another simply for believing in a different theory of time, or explanation for genetic variability, although some may have been tempted and some have colluded with those who did.

But if science will not become a religion, science fiction has already gifted the world one in the form of scientology, much loved by Hollywood actors who find a meaning in it not supplied by personal trainers, truckloads of money and trailers that could house the inhabitants of small countries. C S. Lewis used science fiction to express his own religious beliefs. A religious cult lies at the heart of one of the greatest of science fiction novels, *Dune*, although since Frank Herbert was an American, home of multitudinous cults, perhaps this is not too surprising. Is science fiction itself a kind of religion, asking for the willing suspension of disbelief? Not really. That suspension of disbelief is the price we willingly pay to enter the house of fiction. The pleasure of stepping through that door lies in part in the release it offers from rules and even a confining rationality. When Stephen Hawking went on board the *Enterprise* he was a child again, allowed to play without himself wagging a finger at scientific implausibilities or having others wag a finger at him for the impropriety of being on the bridge of a starship.

If you are British and come, therefore, from a society in which the idea of projecting people from one place to another, at whatever speed, seems largely beyond us, there is something seductive about the idea of teleportation. The one implausibility of *Star Trek* which is bothering is the idea that it will ever be possible to get a decent cup of Earl Grey tea out of a machine, even five centuries from now.

The Poles have a saying, which dates back to the era in which encyclopaedias were rewritten and figures air-brushed from the record: "The hardest thing to predict is the past." Tracking the future is child's play, almost literally so. The past is more difficult to nail down. It forms and reforms, transforms into myth, is reconstituted for present purposes, disappears into the soil to be resuscitated one day as no more than objects whose true function is lost to us. Memory, meanwhile, whispers in our ears, mischievously misleading us as to the path down which we imagine ourselves to have come.

In truth, we exist on a moving frontier where knowledge and the imagination breed new forms. Eden lies far behind. Shakespeare's Prospero was what passed as a scientist on that magic isle where Caliban heard sounds, strange airs which delight and hurt not. He constructed a dream for others to live but in the end he broke his staff, relinquishing a power that had begun to disturb him. There was a cruelty wrapped up in that power and at the last he could no longer live with it. Does knowledge have its moral limits? Is science remote from power? The scientist asks this question and so does the writer of fiction. Scientist and science fiction writer may not climb into bed with one another like Queequeg and Ishmail, or at least not often, but they are shipmates sailing a sea whose depths will never be plumbed and who are driven by knowledge of that fact.

Meanwhile, those who have little time for either rush through space at incredible speed on a planet which rotates so that today in Europe is tomorrow in Australia, and see nothing but the day's delay. To plunge a spade into the earth, or lay an axe to a tree, is to cut down through time just as to explore the human genome is to see back to our origins as the Hubble telescope, image of the future, stares back towards the primal explosion. Science and science fiction, the kissing cousins, are fascinated by what they see and still more by what they will have seen, projecting beyond their deaths and their times as an earnest of their belief in the future which they partly fear and partly rush to embrace.

SMUGGLING SCIENCE ONTO THE PAGE OR STAGE

Carl Djerassi

Carl Djerassi is professor of chemistry at Stanford University (US). He is a member of the US *National Academy of Sciences* and the *American Academy of Arts and Sciences* and other academies around the world. He has received 19 honorary doctorates and such scientific awards as the US *National Medals of Science and Technology*. Professor Djerassi – who moved to the United States from his native Austria during World War II – is a prolific writer. He has produced over 1,200 scientific articles, five novels, two autobiographies and six plays. He describes his creative writing not as fiction but as "science in fiction".

Science fiction is a wildly popular literary genre of long standing with no restrictions as to the plausibility of subject matter. To me – a practising research scientist for half a century who, late in life, decided to find a novel means of communicating with a broader public – this potential freedom to sacrifice scientific reality or plausibility at the altar of literary imagination is a major weakness if one aspires to become a pedagogic smuggler, an aspiration to which I openly admit. I claim such smuggling to be a virtue if one's motive is to bring science to a public that is inherently ascientific or even antiscientific; a public that lowers a curtain labelled "I don't understand science" when they hear the preamble, "let me tell you about my science", and then stop listening. Starting with the more innocent substitute, "let me tell you a story", and then incorporating realistic science and true-to-life scientists into the plot can convert a scientist into an effective smuggler of valuable information. But since plausibility is *de rigueur* if correct scientific information is to be smuggled, I shall offer my definitions of the genres "science-in-fiction" and "science-in-theatre" and then justify their use.

● ● ● Science-in-Fiction

I take it as given that "science fiction" can also be "science-in-fiction". Many writers of science fiction do, of course, base their fantasies on sound knowledge of scientific principles and information. But who will tell the general reader what portion is fact and what is fantasy? When science fiction is meant not only to entertain but also to educate a scientifically illiterate public, the author must make very clear whether and where wild fantasy or flighty prophecy replaces fact and plausibility. Since such clarification is rarely provided in science fiction, a sub-genre, "science-in-fiction", with a clearly defined scope and associated "rules of the game" is more than justified. So what are these? Quite simply, scientific facts and scientific culture should be illustrated in a correct and plausible way. Basing a novel on a perpetual motion machine can be marvelous science fiction, but "science-in-fiction" it is not if my purpose were to inform the reader about thermodynamics.

I certainly did not invent the genre. Sinclair Lewis's *"Arrowsmith"* (which contributed to his winning the Nobel Prize in literature) is an American contribution of the 1920s. C. P. Snow's *"The Search"* in the 1930s, or even better, William Cooper's (a.k.a. Harry S. Huff) *"The Struggles of Albert Woods"* in the 1950s, and Simon Mawer's *"Mendel's Dwarf"* in the 1990s are outstanding British examples. But for every bona fide "science-in-fiction" novel, there are thousands of science fiction counterparts. It would seem unlikely that I coined the term "science-in-fiction", yet I have so far not found any reference to that doubly hyphenated word prior to 1988 when I first used it to describe my own fiction.

But why is science-in-fiction so rare? Is it because so few authors are trained or inclined to write in that genre? Is it because any whiff of "didactic" or "pedagogic" in literature is automatically suspected or criticised? Or is the above definition of science-in-fiction so self-limiting that neither many authors nor readers are attracted to it? What's wrong with following Horace's famous prescription from *Ars Poetica*: "*Lectorem delectando pariterque monendo*" [delighting the reader at the same time as instructing him]? As the author of a tetralogy of science-in-fiction novels and of a trilogy of science-in-theatre plays, I will illustrate my point with some examples.

Scientific Facts in Science-in-Fiction

The popularisation of science, be it in science fiction or science journalism, generally focuses on the actual "science" – i.e. describing <u>what</u> a scientist does or might wish to do. So let me offer two examples from my tetralogy's first volume, *Cantor's Dilemma*, to show how I use science-in-fiction to introduce some entomological titbits. In the first, the scientific hero, Prof. I. Cantor attempts to explain to a non-scientist companion, Paula Curry, why independent replication of scientific experiments is essential. He illustrates the point by reference to a key experiment, performed by his postdoc Stafford in his laboratory, and why failure to repeat an experiment may be due to factors other than incompetence or fraud.

"You mean what Stafford did was wrong?"

"No!" Cantor said sharply. "It doesn't mean that at all. You put it too black and white. It could mean that unknowingly, he has neglected to mention some crucial detail. Or that Krauss's man neglected to follow one.

"But does that happen very often?" Paula was intrigued. She'd always assumed scientific experiments to be unambiguous.

"Absolutely!" Cantor was anxious to make that point. "Let me tell you a true story that will surprise you. Carroll Williams, a distinguished insect biologist at Harvard, once had a Czech postdoctoral fellow, Karel Sláma, come to his laboratory to continue some research he'd started in Prague. He brought with him the bug that he'd studied at home for years, but, lo and behold, Sláma couldn't get the insects to mature and reproduce in Massachusetts, even though he fed them the same diet. You know what it finally came down to?

"The crushed paper on the bottom of the jars in which they kept the insects!" Cantor looked triumphant. "The paper was only supposed to play the role of inert support material – they'd always done that in Prague – but once they eliminated the paper, the insects grew happily in Cambridge." Cantor enjoyed drawing out the story; it made him forget his real problem. "European paper is made from different trees than North American, which is derived from balsam fir. From balsam pulp they isolated what has since been called the 'paper factor', which causes abnormal growth and premature death of the insects. I still remember the conclusion of their otherwise very serious article: the Boston Globe and the Wall Street Journal were inhibitory, whereas Nature and The Times of London were harmless."

Or let me cite an example from the same novel to demonstrate that pheromone sex attractants can at times also be antiaphrodisiacs.

"Celestine arrived at the lecture hall just a few minutes before four. She was impressed to find it nearly full. Her favorite seat – halfway up on the right-hand aisle, suitable for quick departures – was already occupied. She didn't recognize the squatter who'd displaced her, but it wasn't anyone from chemistry. Clearly, Lufkin's title, "One-night Stands among Insects" had brought in the customers. Celestine was curious to see how he'd deliver the goods. She'd never heard Lufkin deliver a research talk.

"Take <u>Lasioglossum zephyrum</u>," he pronounced the words slowly while writing them on the blackboard, "also known as the sweat bee. Some ten years ago, Barrows at the University of Kansas reported that male sweat bees patrol the nests and pounce on females. Such pouncing seems to be promoted by a female odor that can be equated to an aphrodisiac. Barrows was struck by the fact that there were many pounces but very few matings. Aha, thought Barrows, the female sweat bee mates but once.

"Lufkin's eyes swept slowly over the audience. There was hardly a sound as they waited for the next disclosure.

"Along comes Penelope Kukuk from Cornell. Taking some tethered females and exposing them to passing males near the clay bank of a stream – a favorite habitat of sweat bees – Kukuk noted that at least half a dozen males would head for the tethered female. Within a minute or two all but one of the males would depart, leaving a mating pair. Repeating that process with that same tethered, but now deflowered, female showed that she was now unattractive.

"Deflowered tethered female!" How typical of Graham, mused Celestine. I bet half the class is thinking of bondage. But how does one tether a sweat bee?

"You may wonder how one tethers a bee," continued Lufkin as if he were now indulging in a personal dialogue with her. "It's easy. You just tape her by the wings to an applicator stick using Scotch tape. May I have the first slide, please?"

Scientific Culture in Science-in-Fiction

Cantor's Dilemma has so far been translated into eight languages and in the USA has been reprinted annually since 1991[1]. Why? Surely it is

---

1 Djerassi, C., *Cantor's dilemma*, New York, Penguin-USA, 1991.

not because of some untapped reservoir of prurient interest in tethered female sweat bees or other scientific esoterica sprinkled throughout that novel. Rather, it is due to an interesting and initially unexpected shift in my reading public.

As I stated at the outset, I had intended my science-in-fiction to serve as a means of communication with a broader public who would be unlikely to come to my scientific lectures or read my scientific journal publications. But in *Cantor's Dilemma* and the subsequent three volumes (*The Bourbaki Gambit*[2], *Menachem's Seed*[3] and *NO*[4]) of my science-in-fiction tetralogy, I went further than just describing <u>what</u> scientists do. I attempted to draw an accurate picture of the contemporary science scene by describing a tribal culture whose behavioral idiosyncrasies – in other words <u>how</u> they do it – are not only unfamiliar to the outside world but frequently are not even recognised by members of the tribe. Knowledge of how to behave as a scientist – indeed what it means to be a scientist – is generally not taught in courses or books. Scientific "street smarts" – in some respects the soul and baggage of contemporary scientific behaviour – are absorbed by observing the mentor's self-interested concerns with publication practices and priorities, the order of the authors, the choice of the journal, the striving for academic tenure, grantsmanship, *schadenfreude* – even Nobel lust. On their own, budding scientists discover the glass ceiling for women in a male-dominated enterprise, the inherent collegiality of scientific research, and the concurrent brutal competition. Gradually, we accept the white laboratory coat as our cultural uniform.

Since science-in-fiction (in contrast to science fiction) deals not only with real science but also real scientists, it takes an insider to

2 DJERASSI, C., *The Bourbaki Gambit*, New York, Penguin-USA, 1996.
3 DJERASSI, C., *Menachem's Seed*, New York, Penguin-USA, 1998.
4 DJERASSI, C., *NO*, Penguin-USA, New York, 2000.

illuminate some of our more esoteric cultural practices. Unfortunately, the overwhelming majority of working scientists, notably its male members, spend little time on introspection and self-analysis. We are trained to analyse the world around us in exquisite detail and often with amazing insight, but seldom do we apply these skills to an examination of our tribal behaviour. Our 60- to 80-hour working week – itself a manifestation of scientific machismo and obsession – hardly provides the time for such reflective self-examination. I was not very different in that respect until my early sixties when I decided upon a form of auto-psychoanalysis by writing in the guise of fiction about the drive behind creativity and the accompanying posturing and prancing displayed by academics of all ages. Science-in-fiction allowed me a freedom of expression that shame, embarrassment or even fear might have prevented me to use in the form of standard prose. As a result, these novels have become textbooks or recommended reading in many colleges and universities in the types of courses that have proliferated in recent years on the "science-technology-society" and "ethics in research" themes.

To make my point about the tribal behavioural practices of scientists, I shall cite one excerpt from the final volume, "*NO*," of my science-in-fiction tetralogy, where all the characters from the earlier ones reappear. The following conversation between two such women characters and Michael Marletta (in reality, Professor of Biochemistry at the University of California-Berkeley and an example of how I use real scientists with their permission to give my science-in-fiction an extra touch of verisimilitude) deals with publication practices – one of the most esoteric and contentious issues among scientists.

"What have you two been chatting about?" asked Paula.

"Just sociological chitchat," said Marletta. He gestured with his fork toward Celestine. "Now that Professor Price has received tenure," he said with mock formality, "she has started to express quite openly her critical views of our foibles."

Paula arched her eyebrows. "Tell me more. As you will learn this evening, I have become quite interested in the foibles of you all."

Celestine disengaged herself from her aunt's embrace. "I wasn't so much talking about foibles as making an observation. Look at this group." She motioned with her head around the large room. "All of them – no, all of <u>us</u> – are passionately engaged in publishing. It isn't just disseminating scientific information; it's also pushing our names. But look how differently we do it. Max Weiss and Michael always put their names last. They can afford to do that. Everyone knows they're the senior author, irrespective of the number of co-authors."

"Is that true of I.C.?" [Prof. I. Cantor from the first novel, *Cantor's Dilemma*].

"No. I.C. uses alphabetical precedence. A luxury only the As, Bs or Cs can really afford." Celestine's sarcasm was unmistakable. "But what about me? Quite frankly, until I was sure I'd get tenure, I always made sure my name came first. I didn't think I could risk generosity."

"Now, now," admonished Marletta.

"Now nothing," Celestine countered. "What we were talking about, before you came," she turned to Paula, "was the subtlety of how to apportion credit among all the authors. These days, four or more authors are par for the course in any competitive field of chemistry or biology. Having settled who is last, the question now is who comes first."

"Really?" said Paula. "I'd think you'd pick the person who has done most of the work."

"You think that's easy? That's exactly what we've been discussing. Recently, John Scott from Portland published a real first in Science. He had five co-workers, all women – a real harem – but what made it a first was that the first two names listed in the article were marked with an asterisk. Can you guess what the footnote said? 'These authors contributed equally to this manuscript'."

"Brilliant," exclaimed Paula.

"You see?" laughed Marletta.

"Brilliant?" Celestine snorted. "Suppose the first asterisked name had been Smith and the second Price. I would have gone to Scott to point out that in any citation, that article would be referred to either as 'Smith et al' or 'Scott et al.' To me, 'et al' does not mean 'equal'."

"So what would you have had Scott do?"

"Ah," grinned Celestine. "As a first try, I'd have separated the names Smith and Price by an equal sign rather than a comma. But since no editor would allow that, I'd have told him to do it alphabetically."

"You mean I.C.'s system? Why should Smith agree to that when your name starts with a P?"

"Fair enough. That was also Michael's point. So I asked why not toss a coin? And you know what the fair-minded Professor Marletta said?" Celestine poked him lightly with her index finger. "Why don't you tell Paula."

"In my lab, I decide such issues, not the drop of a coin."

"How would you have handled it, Celly?" asked Paula.

"A coin. And that, in my opinion, is one of the differences between a male and a female professor."

"If it is," said Marletta, "it's the only one. Come back in five years and show me in what other respects you behave differently from me."

Clearly, a scientist's ambition and passionate desire for name-recognition is behind such seemingly absurd posturing concerning the order of names in a publication. Even humanists do not have that problem, since generally they publish alone, whereas the days of the noble scientist toiling in solitude are all but gone. Yearning for fame is not unusual – writers, artists, composers, athletes, film and theatre stars share it – but it is mostly scientists who limit such desire for approbation to their professional peers to the virtual exclusion of the general public.

● ● ● Science-in-Theatre

● ● On the Stage

My transmutation late in life from scientist to author of science-in-fiction made me aware how limited the formal written discourse of modern scientists is, since we do not allow the use of dialogue. Was it the limitation of 40 years of monological information transmittal, so typical of scientists, that led me via dialogue-rich science-in-fiction writing to the ultimate dialogic literary form: drama?

By science-in-theatre I refer to plays in which science and scientists do not fulfil primarily a metaphoric function – praiseworthy and attractive as such efforts have been in major plays like Brecht's *Life of Galileo*, Stoppard's *Arcadia* and *Hapgood*, or Dürrenmatt's *The Physicists* – but rather constitute the central focus of the play, as for instance in Frayn's *Copenhagen*. My personal definition of science-in-theatre is even more restricted in requiring also that the science and the behaviour of my scientific characters depicted is real or at least plausible.

So is science-in-theatre simply science-in-fiction presented on the stage rather than within the covers of a book? I prefer to claim that there is more to it. Since the beginning of the Age of Enlightenment, dialogue has essentially disappeared from the <u>written</u> discourse of scientists. Yet dialogue humanises discourse in an important sense and since my objective as a scientist-author is to depict the human foibles and virtues of my tribe's culture, in my recent writing I have turned to the most dialogic form of literature, namely drama.

And while I clearly wish my plays to be seen on the stage, even the most popular plays are only performed occasionally and only in a few locations at any one time. But I believe that science-in-theatre can also be read on its own merits as a tale of science and scientists, which just happens to be presented entirely in dialogic form. My first two science-in-theatre plays, *An Immaculate Misconception*[5], featuring current advances in reproductive medicine, and *Oxygen*[6] (the latter written jointly with Roald Hoffmann) centring around the centenary of the Nobel Prize and the meaning of scientific discovery, have already been translated into nine languages, published in book form in six of them, and broadcast on the BBC World Service. Clearly, they are widely read – be it for pleasure or for instruction. Even the third one, *Calculus*, has already been published in book form in both English and German within a few months of its San Francisco theatrical premiere in April 2003[7].

211

In the Classroom

Having demonstrated – at least to my personal satisfaction – that science-in-theatre is not necessarily the kiss of death in terms of securing professional theatrical performances and an engaged audience as well as readers for book versions of such plays, I have now gone one step further: from the theatre stage to the classroom where new forms of pedagogy should be most welcome. In an attempt to move at least occasionally away from the all-pervasive monologist lecture format in science, I have started on a series of "pedagogic wordplays" for classroom use in lieu of a conventional 50-minute lecture. And rather than expect these to be performed, i.e. learned by heart by actors, students are simply asked to "read" these plays in

---

5 DJERASSI, C., *An Immaculate Misconception*, Londres, Imperial College Press, 2000.
6 DJERASSI, C. & HOFFMANN, R., *Oxygen*, Weinheim, Wiley-VCH, 2001.
7 DJERASSI, C. et PINNER, D., *Newton's Darkness: Two dramatic views*, London, Imperial College Press, 2003.

front of the class with the aid of audiovisuals which are contained in a CD-ROM that is included in the paperback book version of the wordplay. The first of these, *ICSI – Sex in an age of mechanical reproduction*[8] was published in late 2002 in a single paperback containing the English and German texts and has already been performed before hundreds of gymnasium students in Germany as well as college students in the USA. I shall end with an excerpt of the first two minutes from the second of these pedagogic wordplays, *NO* (written with Pierre Laszlo), which appeared in 2003[9] in one slim volume in English, French and German. It may thus provide an additional bonus in language instruction.

Two biochemists – **A** (male) and **B** (female) – sit in a café facing each other on two sides of a table covered with a paper table-cloth and a paper napkin holder. They are drinking coffee and are dressed informally, possibly even in jeans.

A: First… we need money. Otherwise, it's no go!

B: "The love of money is the root of all evil." (Pause). 1 Timothy 6.10. (Laughs). The Bible says so!

A (Dismissive): I'm talking about the <u>need</u> for money… not <u>love</u>. Timothy was no scientist. Or he would've said, "Grant applications are the root of all evil."

B: I hate begging for money… for research!

A: Welcome to the 21st century! (Pause). So let's start on NO.

B: Just a few years ago, this would've been crazy. Fund a collaboration between a chemist and a biologist… on nitric oxide?

A: Especially when most people still confuse ni<u>tric</u> with nit<u>rous</u> oxide… and assume we're interested in laughing gas.

---

8 DJERASSI, C., *ICSI-Sex in the Age of Mechanical Reproduction*, Weinheim, Deutscher Theaterverlag, 2002, with CD.
9 DJERASSI, C. & LASZLO, P., *NO*, Weinheim, Deutscher Theaterverlag, 2003, with CD.

B: And roll their eyes when they learn that <u>nitric</u> oxide… as an industrial gas and environmental pollutant… is toxic! Now we've come full circle… with NO a panacea for God knows how many medical problems. (Pause). Which biological function of NO should we pick?

A: Penile erection.

B: Typical male response.

A: That's hitting below the belt.

B (Jocular): You'll recover.

A: I was thinking of grantsmanship. Penile erection in the title is bound to stand out.

B: Many other catchy applications would also do that.

A: Such as?

B: Migraine, for instance, where NO plays a role. Just thinking of writing our grant application gives me one! (Pause). But let's pretend we're sitting here with another couple… non-scientists… who innocently joined us for coffee.

A: What's that got to do with our research proposal?

B: Assume one of them asks us what we're working on –

A: And we give him or her a tutorial on nitric oxide? (Ostentatiously looks at watch): We're wasting time.

B (Sharply): We're not! It's good discipline… explaining it to the taxpayer…and then putting it into language for a grants committee. Let's give it a try.

A (Reluctantly): Okay… but let's make it snappy. Where do we start?

B: I'd suggest with simple chemistry.

A: Even that will require pencil and paper… or slides… or a blackboard.

B: The eternal plight of chemists unable to explain what they do in simple words! Let's pretend we've only got those napkins.

A: How about the tablecloth? It's also paper.

This *NO* wordplay ends with a RAP, composed and sung by Erik Weiner, which summarises in hip-hop fashion the scientific content of our wordplay. Here is a sampler to demonstrate that scientific content is truly adaptable to any form of communication.

"The N is for Nitric"

(Written and sung by Erik Weiner)

"It's the N to the O, you know, it don't stop
It's time to break it down in the form of hip-hop
The N is for Nitric, the O is for Oxide
It's gotta lotta scientists riled up worldwide

"It's such a hot topic, you saw it today
In **A** and **B** and **VC**'s horseplay
And here we are, ya'll, nearly at the end of it
But before we go we want to recall all the benefits
If you take notes, you can use your pen, so
Once again, the key applications of N-O:

"Number 1, they made it very clear in the first section
That without the N-O, you would get no erection
And without the erection there would be no humpin'
So that's gotta tell you N-O is useful for something."